PRAISE FOR

WHAT DO WOMEN WANT?

"*What Do Women Want?* adds both steam and explosives into the national conversation—or preoccupation—with what it means to be a woman today." —*Vogue*

"Totally engrossing." —*New York* magazine

"In accessible and entertaining prose, *What Do Women Want?* details everything from individual women's fantasies to the search for a 'female Viagra.' More important, though, it represents a complete paradigm shift. . . . This book—how do I put this without sounding hyperbolic? This book should be read by every woman on earth. It should be handed out to pubescent girls right alongside *Our Bodies, Our Selves* and be required course reading for Human Sexuality 101. It is a must-read for any person with even a remote erotic interest in the female gender. It deserves to be listed on bridal registries—gay and straight. It could single-spine-edly replace at least a quarter of the sexual self-help section and the world would be better for it. It is a revelation, a story of redemption. I laughed, I almost cried—with joy. I was turned on, even. You want a female Viagra? This book is as close as we have to it." —Salon

"Bergner lays out the history of this brainwashing and then debunks it in his entertaining new book, *What Do Women Want?* He recaps ingenious studies that have plumbed our desires, including those we deny or hide from ourselves." —*Elle*

"It's everything you wanted to know about sex but didn't know to ask." —*New York Post*

"At last, we have a new perspective on the wilds of female desire, in rousing tableaux, as women, men, sexologists, bonobos, erotic gurus, and many others provide frank, vivid answers to the question that has haunted [us] for far too long: What do women want? The answer will fascinate all."
—Diane Ackerman, author of *A Natural History of Love*

"Daniel Bergner has written a keenly intelligent book about a subject that often exceeds our intelligence: What do women want?" —Gay Talese

"Accessible and informative . . . this page-turning book will have readers questioning some of their most ingrained beliefs about women, men, society, and sex." —*Publishers Weekly*

"Stylishly written . . . an adroit translation of technical material into erudite and entertaining reading." —*Kirkus Reviews*

ALSO BY DANIEL BERGNER

FICTION

Moments of Favor

NONFICTION

The Other Side of Desire:
Four Journeys into the Far Realms of Lust and Longing

In the Land of Magic Soldiers:
A Story of White and Black in West Africa

God of the Rodeo:
The Quest for Redemption in Louisiana's Angola Prison

WHAT DO WOMEN WANT?

ADVENTURES IN THE SCIENCE OF FEMALE DESIRE

DANIEL BERGNER

ecco

An Imprint of HarperCollins*Publishers*

In order to protect the privacy of the women whose sexual lives and personal relationships are described in this book, I have changed names and some minor identifying details. This does not apply to the scientists I've written about or to Shanti Owen, who appears in chapter eight.

Parts of this book originally appeared in different form in the *New York Times Magazine*.

HarperCollins books may be purchased for educational, business, or sales promotional use. For information please e-mail the Special Markets Department at SPsales@harpercollins.com.

A hardcover edition of this book was published in 2013 by Ecco, an imprint of HarperCollins Publishers.

FIRST ECCO PAPERBACK EDITION PUBLISHED 2014.

Designed by Suet Yee Chong

Library of Congress Cataloging-in-Publication Data has been applied for.

ISBN 978-0-06-190609-1

14 15 16 17 18 OV/RRD 10 9 8 7 6 5 4 3 2 1

For Georgia

CONTENTS

WHAT DO WOMEN WANT?

CHAPTER ONE

Animals

On the subject of women and sex, Meredith Chivers was out to obliterate the civilized world. The social conventions, the lists of sins, all the intangible influences needed to go. "I've spent a lot of time," she said, "attempting to get back in my head to what life was like for proto-humans."

When Chivers and I first met seven years ago, she was in her mid-thirties. She wore high-heeled black boots that laced up almost to her knees and skinny, stylish rectangular glasses. Her blond hair fell over a scoop-necked black top. She was a young but distinguished scientist in a discipline whose name, sexology, sounds something like a joke, a mismatching of prefix and suffix, of the base and the erudite. Yet the matching is in earnest—the ambitions of the field have always been grand. And Chivers's dreams were no different. She hoped to peer into the workings of the psyche, to see somehow past the consequences of culture, of nurture, of all that is learned, and to apprehend a piece of women's primal and essential selves: a fundamental set of sexual truths that exist—inherently—at the core.

Men are animals. On matters of eros, we accept this as a kind of psychological axiom. Men are tamed by society, kept, for the most part, between boundaries, yet the subduing isn't so complete as to hide their natural state, which announces itself in endless ways—through pornography, through promiscuity, through the infinity of gazes directed at infinite passing bodies of desire—and which is affirmed by countless lessons of popular science: that men's minds are easily commandeered by the lower, less advanced neural regions of the brain; that men are programmed by evolutionary forces to be pitched inescapably into lust by the sight of certain physical qualities or proportions, like the .7 waist-to-hip ratio in women that seems to inflame heterosexual males all over the globe, from America to Guinea-Bissau; that men are mandated, again by the dictates of evolution, to increase the odds that their genes will survive in perpetuity and hence that they are compelled to spread their seed, to crave as many .7's as possible.

But why don't we say that women, too, are animals? Chivers was trying to discover animal realities.

She carried out her research in a series of cities, in Evanston, Illinois, which sits right next to Chicago, in Toronto, and most recently in Kingston, Ontario, which feels utterly on its own, tiny, and fragile. The Kingston airport is barely more than a hangar. Kingston's pale stone architecture has a thick, appealing solidity, yet it doesn't chase away the sense that the little downtown area, on the frigid spot where Lake Ontario spills into the Saint Lawrence River, isn't much more formidable than when it was founded as a French fur-trading post in the seventeenth century. Kingston is the home of Queen's University, a sprawling and esteemed institution of learning, where Chivers was a psychology professor, but the city is stark and scant enough that it is easy to imagine an earlier emptiness, the buildings gone, the pavement gone, almost nothing there except evergreens and snow.

And this seemed fitting to me when I visited her there. Because to reach the insight she wanted, she needed to do more than strip away societal codes; she needed to get rid of all the streets, all the physical as well as the incorporeal structures that have their effects on the conscious and the unconscious; she needed to re-create some pure, primordial situation, so that she could declare, This is what lies at the heart of women's sexuality.

Plainly, she wasn't going to be able to establish such conditions for her studies. Almost surely, for that matter, such pure conditions never existed, because proto-humans, our forehead-deficient *Homo heidelbergensis* and *Homo rhodesiensis* ancestors of some hundreds of thousands of years ago, had proto-cultures. But what she possessed was a plethysmograph: a miniature bulb and light sensor that you place inside the vagina.

This is what her female subjects did as they sat on a brown leatherette La-Z-Boy chair in her small, dimly lit lab in Toronto, where she first told me about her experiments. Semireclining on the La-Z-Boy, each subject watched an array of porn on an old, bulky computer monitor. The two-inch-long glassine tube of the plethysmograph beams light against the vaginal walls and reads the illumination that bounces back. In this way, it measures the blood flow to the vagina. Surges of blood stir a process called vaginal transudation, the seeping of moisture through the cells of the canal's lining. So, indirectly, the plethysmograph gauges vaginal wetness. It was a way to get past the obfuscations of the mind, the interference of the brain's repressive upper regions, and to find out, at a primitive level, what turns women on.

As they enrolled in the study, Chivers's subjects had identified themselves as straight or lesbian. This is what all of them saw:

A lush-bodied woman lay back beneath her lover on a green army blanket in the woods. His hair was cropped, his shoulders hulking. He propped his torso on rigid arms and slid inside her. She lifted her thighs and enwrapped him with her calves. The pace of his thrusting quickened, the muscles of his buttocks rippled, her fingers spread and seized his triceps.

After each ninety-second clip of porn, the subjects watched a video that sent the plethysmograph's readings back to a baseline state. The camera scanned jagged mountains and rested on a parched plateau.

Then a man walked naked on a beach. His back formed a V, and ridges of muscle angled toward his groin above his taut thighs. He flung a stone into the surf. His chest was massive. So were his buttocks, without a hint of fat. He strode along a rock precipice. His penis, relaxed, slung from side to side. He tossed another stone and stretched his spectacular back.

A slender woman with a soft, oval face and dark, curly hair sat on the lip of a large tub. Her skin was tan, her areolas dark. Another woman rose from the water, her soaked blond hair raked behind her ears. She pressed her face between the brunette's thighs and whisked with her tongue.

On his knees an unshaven man mouthed a sizeable penis that rose below a sheer, muscled stomach.

A woman with long black hair leaned forward on the arm of a lounge chair, her smooth buttocks elevated. Then she settled her light brown body onto the white upholstery. Her legs were long, her breasts full, high. She licked her fingertips and stroked her clitoris. She pulled her spread knees up. She handled one breast. Her hips began to grind and lift.

A man drove himself into the ass of another man, who let out a grateful moan; a woman scissored her legs in a solitary session of nude calisthenics; a bespectacled, sculpted man lay on his back and masturbated; a man slipped a woman's black thong over her thighs and began with his tongue; a woman straddled another woman who wore a strap-on.

Then a pair of bonobos—a species of ape—strolled through a grassy field, the male's reedy, pig-colored erection on view. Abruptly, the female splayed herself, her back on the ground and legs in the air. While her mate thrust into her, his rhythm furious, she threw her hands above her head, as if in total erotic surrender.

Sitting on the leatherette chair, Chivers's subjects, straight and lesbian, were turned on right away by all of it, including the copulating apes. To stare at the data amassed by the plethysmograph was to confront a vision of anarchic arousal.

This was my initial glimpse of sexology's strivings after female desire. Chivers's husband, a psychologist whose thinking I'd sought out for another book about sex, introduced us. And soon

I was learning not only from Chivers but from many of the researchers she called a "gathering critical mass" of female scientists who were set on puzzling out the ways of eros in women. There was Marta Meana with her high-tech eye-tracker and Lisa Diamond with her low-tech, long-term studies of women's erotic existences and Terri Fisher with her fake polygraph machine. Men, too, were part of the project. There was Kim Wallen with his monkeys and Jim Pfaus with his rats. There was Adriaan Tuiten with his genetic screening and his specially designed aphrodisiacs, Lybrido and Lybridos, that were headed to the Food and Drug Administration for approval.

And while they tutored me in their labs and animal observatories, I was listening as well to numberless everyday women who shared their yearnings and their bewilderment, who explained what they could—and couldn't—understand about their sexuality. Some of their stories are laced throughout these pages. There was Isabel, who, in her early thirties, was tormented by a basic question: whether she should marry the handsome and adoring boyfriend she had once—but no longer—desired. Every so often, when they stood at a bar, she told him, "Kiss me like we've never met before." She felt a reverberation, terribly faint, instantly fading. It mocked her, teaching her repeatedly: better not to make requests like that. "I'm not even thirty-five," she said to me. "That tingling—I don't get to feel that anymore?" And there was Wendy, who, ten years older than Isabel, had signed up for the Lybrido and Lybridos trials, to see if an experimental pill could restore some of the wanting that had once overtaken her with her husband, the father of her two children.

Others I interviewed—like Cheryl, who was slowly, deliberately reclaiming her capacity for lust after disfiguring cancer surgery, or Emma, who wanted our conversation to start at the strip club where she'd made her living a decade ago—don't appear in these chapters but invisibly inform them. I interviewed and interviewed and interviewed, hoping for yet more sight lines, and in the end, recent science and women's voices left me with pointed lessons:

That women's desire—its inherent range and innate power—is an underestimated and constrained force, even in our times, when all can seem so sexually inundated, so far beyond restriction.

That despite the notions our culture continues to imbue, this force is not, for the most part, sparked or sustained by emotional intimacy and safety, as Marta Meana would stress both in front of her eye-tracker and beside a casino stage.

And that one of our most comforting assumptions, soothing perhaps above all to men but clung to by both sexes, that female eros is much better made for monogamy than the male libido, is scarcely more than a fairy tale.

Monogamy is among our culture's most treasured and entrenched ideals. We may doubt the standard, wondering if it is misguided, and we may fail to uphold it, but still we look to it as to something reassuring and simply right. It defines who we aim to be romantically; it dictates the shape of our families, or at least it dictates our domestic dreams; it molds our beliefs about what it means to be a good parent. Monogamy is—or we feel that it is—part of the crucial stitching that keeps our society together, that prevents all from unraveling.

Women are supposed to be the standard's more natural

allies, caretakers, defenders, their sexual beings more suited, biologically, to faithfulness. We hold tight to the fairy tale. We hold on with the help of evolutionary psychology, a discipline whose central sexual theory comparing women and men—a theory that is thinly supported—permeates our consciousness and calms our fears. And meanwhile, pharmaceutical companies search for a drug, a drug for women, that will serve as monogamy's cure.

CHAPTER TWO

Bodies and Minds

Chivers traced her love of collecting data back to her father, a Canadian Air Force colonel. With a master's in the field of human factors engineering, he created efficient cockpits for fighter jets; he studied reaction times to signals and how best to arrange a plane's controls. He taught her a reverence for the empirical. He plucked up a rock and told her about geological formations; he uncovered earthworms and talked about the aeration of soil. When the weekly TV section arrived with the newspaper, she underlined all the science shows. For her pet hamsters, she built mazes out of cardboard boxes. She settled on an optimal reward—the smell of peanut butter, she discovered, was too pervasive and confusing, so she used vegetables—and ran experiments to learn whether the nocturnal rodents functioned more effectively and found a route to the food faster at night.

Down in her father's basement workshop, she learned to build under his watch and made a fridge with tiny wire hinges and a horse stable to go with the dollhouse he fabricated. She was entranced by the way things—inanimate and animate—fit together and operated; by college she was studying neuroscience, devoting herself to biophysics and biochemistry, when a friend suggested she enroll in something easy, a sexuality class. Six hundred students filled the lectures. One day the professor was showing slides. A vulva appeared. The ridges and folds of female genitalia, in tight close-up, took over the screen. Disgust consumed the hall, a massive expulsion of "Eeew!" that Chivers heard mostly from the women. A close-up of a penis caused no horror, no gasp, from anyone.

Back in high school, for a group of male classmates, Chivers had sketched the vulva's anatomy, a map to help the boys in finding the clitoris. Now, surrounded by the women's voluble wincing, she thought, This is the way you feel about your own bodies?

After the lecture course, she enrolled in a sexuality seminar. She gave a presentation on women's problems with orgasm; she played a video of a woman in her sixties talking about a new partner, a late awakening. She led an electric discussion and left the room elated, but she couldn't conceive of any career dealing directly with sex, besides being a sex therapist, which she didn't want. Sticking with neuropsychology, she wound up doing a thesis experiment that added to fledgling evidence: that homosexual men perform less well than heterosexuals on a type of test involving three-dimensional shapes, just as females, on average, perform less well than males.

This bit of undergraduate research wasn't very politic. It fell within an area of science that is fiercely debated, mostly because of its signs that there are certain differences in intelligence between women and men due not to culture but to genes. Yet Chivers didn't much care about the politics; she was gazing at an intriguing nexus: between gender (the skill discrepancies between women and men at rotating three-dimensional shapes in their minds), desire (the similar discrepancies between gays and straights), and aspects of neurology that might well be innate. After graduation, she begged her way into an assistant's job at the Toronto lab that would later, after she got her doctorate, contain her cramped chamber with its La-Z-Boy and plethysmograph; it is part of one of Canada's most prestigious psychiatric teaching hospitals. When she arrived at the age of twenty-two, she was the only woman on the floor. Male sexuality was the sole focus of the science being done, and one day she asked the oldest researcher, Kurt Freund, an eighty-one-year-old icon in sexology, why he never turned his attention to women.

Bald, with a bladelike nose and oversized ears that seemed to be sly instruments of detection, Freund, a Czechoslovakian psychiatrist, had been hired by the Czech military half a century earlier to catch conscripts who were trying to escape service by pretending to be homosexual. He'd developed a male version of the plethysmograph. This was long before a female equivalent existed. A glass tube was placed over the penis with an airtight seal around the base of the shaft. Images were shown. A gauge determined air pressure and marked the swelling. If, with a Czech draftee, the pressure didn't rise when Freund showed provocative pictures of young men, the conscript was headed into the army.

Freund didn't make a career out of hunting homosexuals. Early on, he tried to cure gays through psychoanalysis; eventually he called in his patients and gave their money back. Arguing that homosexuality arose from prenatal biology rather than upbringing, and insisting that it could not be treated, he fought Czech laws that criminalized gay sex. After he fled communist rule and settled in Toronto, his vision of male sexual orientation as permanent—and being gay as nothing like a sickness—helped to convince the American Psychiatric Association, in 1973, to strike homosexuality from its list of mental disorders.

Like all the researchers at the Toronto lab, Freund highlighted the inborn scripting of desire. Nurture was in constant interplay with nature, but it wasn't a fifty-fifty partnership. Answering Chivers, he asked a question of his own: "How am I to know what it is to be a woman? Who am I to study women when I am a man?" His words put her on the far side of a divide—a chasm, in his view. For her, they laid down a challenge. There were experiments to be constructed, data to be compiled, deductions to be distilled, results to be replicated. She imagined one day drawing a map that would capture female eros. "I feel like a pioneer at the edge of a giant forest," she said, when we first spoke. "There's a path leading in, but it isn't much."

In her sense of quest, there were echoes of Sigmund Freud, of his words to Marie Bonaparte almost a century ago. A great-grandniece of Napoleon, she was one of Freud's psychoanalytic disciples. "The great question that has never been answered," he told her, "and which I have not yet been able to answer, despite my thirty years of research into the feminine soul, is, What does a woman want?"

* * *

While they watched the clips of erotica, Chivers's subjects didn't just wear a plethysmograph; they also held a keypad. On this, they rated their own feelings of arousal. So Chivers had physiological and self-reported—objective and subjective—scores. They hardly matched at all. All was discord. And this dissonance resonated loudly with findings from other researchers.

Women with women, men with men, men with women, lone men or women masturbating—Chivers's objective numbers, tracking what's technically called vaginal pulse amplitude, soared no matter who was on the screen and regardless of what they were doing, to each other, to themselves. Lust was catalyzed; blood flow spiked; capillaries throbbed indiscriminately. The strength of the pulsings did hold a few distinctions, variations in degree, one of them curious: the humping bonobos didn't spur as much blood as the human porn, but with an odd exception. Among all women, straight as well as gay, the chiseled man ambling alone on the beach—an Adonis, nothing less—lost out to the fornicating apes. What to make of such strangeness?

There was some further discrimination on the part of the lesbians. Over the series of studies Chivers did—to be sure her data were no fluke—they were a little selective; amplitude leapt more during videos starring women. Yet the lesbians' blood rushed hard during scenes of gay male porn. When Chivers analyzed the evidence, transmitted from vaginal membranes to sensor to software, when she set it out in graphs of vertical bars, the female libido looked omnivorous.

The keypad contradicted the plethysmograph, contradicted

it entirely. Minds denied bodies. The self-reports announced indifference to the bonobos. But that was only for starters. When the films were of women touching themselves or enmeshed with each other, the straight subjects said they were a lot less excited than their genitals declared. During the segments of gay male sex, the ratings of the heterosexual women were even more muted—even less linked to what was going on between their legs. Chivers was staring at an objective and subjective divide, too, in the data from the lesbians: low keypad scores whenever men were having sex or masturbating in the films.

She put heterosexual and homosexual males through the same procedure. Strapped to their type of plethysmograph, their genitals spoke in ways not at all like the women's—they responded in predictable patterns she labeled "category specific." The straight men did swell slightly as they watched men masturbating and slightly more as they stared at men together, but this was dwarfed by their physiological arousal when the films featured women alone, women with men, and, above all, women with women. Category specific applied still more to the gay males. Their readings jumped when men masturbated, rocketed when men had sex with men, and climbed, though less steeply, when the clips showed men with women. For them, the plethysmograph rested close to dead when women owned the screen.

As for the bonobos, any thought that something acutely primitive in male sexuality would be roused by the mounting animals proved wrong. The genitals of both gay and straight men reacted to these primates the same way they did to the landscapes, to the pannings of mountains and plateaus. And with the men, the objective and subjective were in sync. Bodies and minds told the same story.

How to explain the conflict between what the women claimed and what their genitals said? Plausible reasons swirled. Anatomy, Chivers thought, might be one factor. Penises extend, press against clothes. Visibly they shrink and shrivel. Boys grow up with a perpetual awareness; male minds are used to being fed information from their groins. A sexual loop between body and cognition, each affecting the other, develops; it runs fast and smooth. For women, more covert architecture might make the messages less clear, easier to miss.

But were the women either consciously diminishing or unconsciously blocking out the fact that a vast scope of things stoked them—stoked them instantly—toward lust?

The discord within Chivers's readings converged with the results of a study done by Terri Fisher, a psychologist at Ohio State University, who asked two hundred female and male undergraduates to complete a questionnaire dealing with masturbation and the use of porn. The subjects were split into groups and wrote their answers under three different conditions: either they were instructed to hand the finished questionnaire to a fellow college student, who waited just beyond an open door and was able to watch the subjects work; or they were given explicit assurances that their answers would be kept anonymous; or they were hooked up to a fake polygraph machine, with bogus electrodes taped to their hands, forearms, and necks.

The male replies were about the same under each of the three conditions, but for the females the circumstances were crucial. Many of the women in the first group—the ones who could well have worried that another student would see their answers—said they'd never masturbated, never checked out anything X-rated. The women who were told they would have strict confidentiality answered yes a lot more. And the women who thought they were wired to a lie detector replied almost identically to the men.

Because of the way the questions were phrased—somewhat delicately, without requiring precise numbers, Fisher told me, in deference to the conservative undertone she sensed on her satellite campus—the study couldn't pinpoint rates of masturbation or porn use; yet, she went on, it left no doubt as to the constraints most women feel about acknowledging the intensity of their libidos. When Fisher employed the same three conditions and asked women how many sexual partners they'd had, subjects in the first group gave answers 70 percent lower than women wearing the phony electrodes. Diligently, she ran this part of the experiment a second time, with three hundred new participants. The women who thought they were being polygraphed not only reported more partners than the rest of the female subjects, they also—unlike their female counterparts—gave numbers a good deal higher than the men.

This kind of conscious suppression could well have distorted the self-reports of Chivers's straight women, but had it insinuated itself with the lesbians? Many of them might have adopted a stance of defiance about their sexuality—wouldn't this have lessened any impulse toward lying? Maybe, though with these women another sort of restraint could have been at work: the need for fidelity to their orientation, their minority identity.

Fisher's research pointed to willful denial. Yet, Chivers believed, something more subtle had to be at play. In journals she found glimmers of evidence—unconfirmed, insubstantial, like so much that she wished she could rely on, build on, as she attempted to assemble sexual truth—that women are less connected to, less cognizant of, the sensations of their bodies than men, not just erotically but in other ways. Was there some type of neural filter between women's bodies and the realms of consciousness in the brain? Something tenuous about the pathways? Was this especially the case with sexual signals? Was this a product of genetic or societal codes? Were girls and women somehow taught to keep a psychic distance from their physical selves? Deep into our seven-year conversation, Chivers spoke bluntly about the congenital and the cultural, about nature and nurture and women's libidos. For a long while, though, she made no pronouncements. Her scientific intentions were aggressive, the stripping away of the societal, the isolation of the inherent. But she had a researcher's caution, an empiricist's reserve, a reluctance to declaim more than the data could support.

Fisher, meanwhile, was emphatic about the contortions imposed, the compressions enforced. "Being a human who is sexual," she said, "who is *allowed* to be sexual, is a freedom accorded by society much more readily to males than to females." Her lie detector was unequivocal.

Rebecca was a forty-two-year-old elementary school music teacher with three children. One afternoon, on the computer she shared with her husband, she discovered a picture of a woman who was plainly his lover. In all sorts of ways, this was devastating. There was the difference in age between the two women, clear to Rebecca immediately. More particularly and insidiously, there were the woman's breasts, exposed in the photograph and, in Rebecca's eyes, significantly superior to her own, which had shrunk, she was sure, more than most do from nursing. And then there was her sense—instantaneous—that her husband wanted the photo to be found and the affair to be found out, because he hadn't had the courage to end the marriage and move in with the woman—who was blowing a whimsical kiss from the screen—without some mayhem to camouflage the long premeditation of his escape. Obeying a therapist's advice, Rebecca tried not to beg her husband to stay. She lobbied through friends. She gave her husband a book about seeking spiritual fulfillment instead of chasing new love. But within weeks, she was a single mother who spent a good amount of time in front of the computer, comparing herself to the seminude picture, which she'd forwarded to her own email address.

Rebecca—who was among the women I spent my time talking with, questioning relentlessly—had a talent for self-disparagement. This encompassed everything from her body to her career. How had she wound up teaching flute and clarinet to fourth-graders and never performing herself except during the intermissions at her students' recitals? And how, she wondered further, had she managed to corner herself into this marmish existence in, of all places, Portland, Oregon, America's city of hipsters?

Yet her skill at self-denigration was matched by a fiery resilience. Increasingly, on her screen the image of the twenty-nine-year-old girlfriend was replaced by the home site of an Internet dating service.

Gradually she had some dates. And gradually there was a man she saw as attractive and felt was kind, and—even before they slept together—she confided, over dinner in a Thai restaurant, something that had taken her fourteen years to tell her husband. She wanted to have a threesome with a woman. The discord and dissembling that ran through Chivers's and Fisher's findings weren't her issues. Why she'd waited so long to raise her desire with her husband she wasn't certain. Yes, some shyness was involved, but she guessed it had more to do with a hunch that turned out to be prescient: he didn't show any interest. Probably, she thought, this was because having another woman in their bed would have made a glaring reality out of his lack of interest in Rebecca herself. In any case, her date agreed that a threesome would be good. They abandoned the topic there, began sleeping together, and returned to the subject a few months later. She said that she would leave the arrangements to him.

He asked whether she had any criteria. She'd never been with a woman in a threesome or in any other way. Her wishes were specific. Hair color different from her own. Not too tall. In decent shape. White or Latina. And—a factor she'd been fixated on for years—large breasts. C-cups, at a minimum, as long as they were real.

She and her boyfriend joked that she was as male as any caricature of a man. Because he'd never done anything like this before, it took him a while, but eventually he presented her with possibilities. He showed her a photograph from a casual connections site, a woman Rebecca found herself fantasizing about right away. But the emailing with this candidate flickered, and the chance faded. They debated whether to hire an escort. Periodically during this process of false starts, Rebecca was seized by fear: what if the woman saw her as old, repellent? But her boyfriend reassured her, and her desire was louder in her mind than her worry, and as they shifted toward the idea of renting someone, she reminded herself that her own attractiveness simply wasn't supposed to matter.

At last, with a babysitter taking care of her children, they waited at his apartment for the arrival of an escort he'd chosen from row after row of thumbnail images online. Wanting to be welcoming hosts and to soften the prostitution-like aspects of hiring a prostitute, they had lit tea candles and chilled a nice bottle of wine. When the escort rang the doorbell, though, and when Rebecca and her boyfriend glanced out the living room window, the harsher qualities of the situation became more difficult to ignore.

Despite her high price, the woman was homely and built along the lines of a lumpy block. Rebecca whispered to her boyfriend that maybe the homeliness was due to the glare of his porch light, that all would be okay once they opened the door and began. She felt relieved, meanwhile, that she wouldn't have to be concerned about her own looks. But when they opened the door and the escort stepped quietly, even timidly, into the vestibule, with the manner more of a housemaid than a call girl, the trouble did not improve. The woman appeared to be around ten years older than Rebecca. And now Rebecca was calculating at rapid speed whether she should and could go through with this to spare the prostitute's feelings, so that the problem was no longer how to soften the exploitation of a body but how to avoid letting this woman know that her body was unexploitable.

Rebecca all but prayed that her boyfriend would somehow solve everything. He told the escort that Rebecca had suddenly come down with something, that she wasn't up to it, an excuse that sounded about as convincing as her fourth-graders' explanations for not practicing their instruments, though the woman, who smiled graciously, seemed to accept the reason or, either way, to be grateful not to have to perform. He gave her some minor cash for her time, and Rebecca said good-bye in sweet tones, and she and her boyfriend went upstairs to click on his computer and stare for a few moments in befuddlement at the immense disparity between the picture and the person and to discuss the mystery of how other customers had handled this difference and whether it was a common dilemma in securing an escort and how you were supposed to prevent this from happening. "I think you just have to spend more," Rebecca said.

So they did. The second woman was pretty and young. She, too, was at odds with her picture, but not drastically, and Rebecca immersed herself in the escort's breasts, in her thighs, in her lips, in all the parts that had been paid for, lost herself in the textures and sights and smells, and was nearly euphoric afterward, both because, after years and years of yearning, she'd broken through the range of barriers that stood between her and another woman's body and lost her virginity in this sense and because, leaving the breakthrough aside, there'd been such pleasure in having, among other things, the prostitute's nipples in her mouth.

When Rebecca and I talked, she said that while she hoped for another threesome with a woman soon and might like to have a woman alone, she didn't much think of herself as a lesbian nor really as bisexual. She had no doubt that she preferred the romantic company of men. She fantasized mostly about men, was still happily with the same boyfriend, and definitely wouldn't want to replace him with a woman. I described Chivers's plethysmographic readings and asked for her thoughts.

The results didn't mean that women secretly want to have sex with bonobos, she began, laughing. And it might not be right to label most women as bi, even if lots of women, like her, did wish to have sex with women or would if they permitted themselves to know it. "It's hard to find the right words," she said. "The phrase that keeps coming into my head is that it's like a pregnancy of wanting. Pregnancy's not a good word—because it means pregnancy. It's that it's always there. Or always ready. And so much can set it off. Things you actually want and things you don't. Pregnant. Full. The pregnancy of women's desire. That's the best I can do."

Stranger. Close friend. Lover of long-standing.

This was the focus of a new experiment Chivers was finishing during one of my visits. The results made her pulse quicken.

It didn't race all that often. The daily labors of her research were painstaking; her office in Kingston was about as spare as a monk's cell. The cinder-block walls were nearly bare. Taped above her desk were a few splotches of purple and green painted by her toddler son. On the opposite wall was a small photographic triptych she'd taken of stone carvings at an Indian temple. A man, in the first image, had sex with a mare while another masturbated; a couple tongued each other's genitals in the middle picture; in the last photo seven human figures were lost in orgiastic heat. Yet for all its drama, the triptych was miniature enough to overlook. The cinder block dominated; there was minimal distraction; she wanted it like this. She could imagine herself surrounded by what she was venturing into, the forest of female desire.

One morning at her metal desk, with a flat November light making its way through her window, she bent over her laptop, poring through plethysmographic readings she'd collected in her latest study. Her eyes tracked a jagged red line that ran across the screen, a line that traced one subject's blood flow, second by second by second. Before Chivers could use a computer program to take the data and arrange them in a meaningful form, she needed to eliminate errant points, moments when a subject had probably shifted in her chair, generating a slight pelvic contraction and jarring the plethysmograph, which could, in turn, cause a jolt in the readings and skew the overall results. Slowly, she scanned the line with all its cramped zigs and zags, searching for spots where the unusual height of a peak relative to the ridges beside it told her that arousal wasn't at play, that an interval was irrelevant to her study. She highlighted and deleted one tiny aberrant section, then continued squinting. She would search in this way for about two hours in preparing the data of a single subject. "I'm going blind," she said, as she stared at another suspicious crest.

She was thrilled, though, with what her experiments were

uncovering—and thrilled to belong within the "gathering critical mass," an unprecedented female effort. The discipline of sexology, which was founded in the late nineteenth century, had always been a male domain. Even now, women made up less than a third of the membership in the field's most eminent organization, the International Academy of Sex Research, and less than a third of the editorial board—on which Chivers served—of the Academy's journal. So female eros hadn't been examined with nearly as much energy as it might have been. And one of Chivers's heroes, one of the older women in the field, Julia Heiman, the director of the Kinsey Institute at Indiana University, told me that, in addition, sexology had for many decades devoted itself more to documenting behavior than to looking into the feelings, like lust, that lie underneath. Alfred Kinsey's work at midcentury, she said, didn't reveal all that much about desire. He had started his career as an entomologist, cataloguing species of wasps; he was wary of delving into emotion. William Masters and Virginia Johnson, filming hundreds of subjects having sex in their lab, drew conclusions that concentrated on function rather than craving. It wasn't until the seventies that sexologists began zeroing in on what women want rather than what women do. And then AIDS engulfed the attention of the discipline. Prevention became everything. Only in the late nineties did full-scale exploration of desire start again.

In her new experiment, Chivers played pornographic audiotapes, instead of videos, for straight female subjects. Ever meticulous, always intent on duplicating her results from alternate angles, she wanted to know, partly, whether spoken stories would somehow have a different effect on the blood, on the mind, on the gap between plethysmograph and keypad. "You meet the real estate agent outside the building. He shows you the empty apartment. . . ." "You notice a woman wearing a clinging black dress, watching you. . . . She follows you. She closes the door and locks it. . . ." The scenes her subjects heard varied not only by whether they featured a man or a woman in the seductive role but by whether the scenario involved someone unknown, or known well as a friend, or known long as a lover. There was the female friend dripping in her bathing suit at the side of the pool. There was the male roommate; there was the female stranger in the locker room. All were depicted as physically alluring, and all the salient details were kept equivalent: the pacing of the ninety-second narratives, the abrupt hardness of the cocks, the swelling of the nipples.

Once again, when all was analyzed, the gap was dramatic: the subjects reported being much more turned on by the scenes starring males than by those with females; the plethysmograph contradicted them. Chivers was pleased by the confirmation. But this time, it was something else that excited her.

Genital blood throbbed when the tapes described X-rated episodes with female friends—but the throbbing for female strangers was twice as powerful. The broad-chested male friends were deadening; with them, vaginal pulse almost flatlined. The male strangers stirred eight times more blood.

Chivers's subjects maintained that the strangers aroused them least of all the men they heard about. The plethysmograph said the opposite. Longtime lovers, male or female, were edged out by the unknown men or women—even though the lovers were dreams, perfect. Sex with strangers delivered a blood storm.

This didn't fit well with the societal assumption that female sexuality thrives on emotional connection, on established intimacy, on feelings of safety. Instead, the erotic might run best on something raw. This idea wasn't completely new, but it tended to be offered as the exception rather than the rule: the raw was important to few women; it was the material of only intermittent fantasy for most. Here was systematic evidence to the contrary, the suggestion of a new, unvarnished norm.

Chivers's work emphasized discord not only between bodies and minds but between realities and expectations, and around her, other researchers, too, were calling conventions into doubt. One was the old notion that women's sexuality is innately less visual than men's. Kim Wallen, an Emory University psychology professor whose hordes of rhesus monkeys I visited between my tutorials with Chivers, collaborated with Heather Rupp, his former student and a sexologist at the Kinsey Institute, in showing erotic photographs to male and female subjects. They used viewing time, down to the thousandth of a second, to measure level of interest. The women gazed at the porn no less long than the men. It seemed they were just as riveted.

Terri Conley, a psychologist at the University of Michigan, had been dwelling for years on a series of studies, done over the past four decades, affirming repeatedly that men welcome casual sex while women, for the most part, don't. In two of these experiments, males and females—"of average attractiveness," as the researchers described them, and around twenty-two years old—were sent out onto a college campus to proposition two hundred members of the opposite sex. Either they asked for a date, or they asked, "Would you go to bed with me tonight?" About the same percentage of men and women—50 percent or so—answered yes to the date. But close to three-quarters of the male responders and none of the females said yes to bed. The data had been used often to argue not only a vast but an intrinsic difference in the desires of men and women. Conley created a questionnaire to look at the topic in another way.

Her two hundred college-aged subjects, all of them heterosexual, were asked to imagine scenarios like this: "You are fortunate enough to be able to spend your winter vacation in Los Angeles. One day, about a week into your stay, you decide to visit a trendy café in Malibu that overlooks the ocean. As you are sipping your drink, you look over and notice that the actor Johnny Depp is just a few tables away. You can hardly believe your eyes! Still more amazing, he catches your eye and then approaches you. . . ."

"Would you go to bed with me tonight?" Depp asked the female subjects. So did Brad Pitt and Donald Trump. The males were approached by Angelina Jolie, Christie Brinkley (chosen by Conley because she wondered whether at fifty-something a woman's age would undercut her appeal despite her extreme beauty—it didn't seem to), and Roseanne Barr. The experiment stripped away the social expectations, as well as the physical risks, that auger against a woman consenting to have sex with a stranger. Conley's setup left only fantasy, frequently a clearer window into desire. The subjects scored how they felt about the propositions. The women were just as avid about saying yes to Depp and Pitt as the men were with Jolie and Brinkley; the women were just as hungry, impulsive, impelled. Trump was dismissed with as much distaste as Barr.

Chivers, when she moved on to her next study, found something that complicated what she'd been seeing. But it also crystalized the raw portrait of female lust that was emerging in her work and the research of her colleagues.

A set of straight women looked at pictures of male and female genitalia. There were four kinds of photos: one with a dangling penis; another with a taut erection; a third with a demure vulva half-concealed by coy thighs. The fourth was a "full-on crotch shot," Chivers said, with typical wry humor, of a woman with spread legs. In all four, the genitalia were tightly framed, mostly disembodied; there was little else to be seen. This time, the subjects' blood wasn't indiscriminate. It rushed much, much more when an erection occupied the screen than when any of the other images were on the monitor. Paradoxically, here was objective evidence that women were categorical after all. And this jibed with what Rebecca had said, that she didn't quite think of herself as bisexual, that she felt an inescapable preference for men even as she harbored plenty of lust for women. It resonated, too, with the faint reactions of Chivers's earlier subjects when the Adonis with the slack penis walked along the shore. It seemed that the visible slackness had nullified the rest of his impressive body. More than anything, though, as an isolated, rigid phallus filled vaginal blood vessels and sent the red line of the plethysmograph high, niceties vanished, conventions cracked; female desire was, at base, nothing if not animal.

CHAPTER THREE

The Sexual Fable of Evolutionary Science

The history of sexuality, and perhaps above all the history of women's sexuality, is a discipline of shards. And it is men, with rare exceptions, whose recorded words form the fragments we have of ancient and medieval and early modern ideas about female eros. Such glimpses are worth only so much. But what can be said about these fragments is that they add up to a particular sort of balance—or imbalance—between an acceptance and even a celebration of desire and drive on the one hand and, on the other, an overriding fear.

A woman in the Bible's Song of Songs:

I sleep, but my heart is awake
I hear my love knocking.
"Open to me, my sister, my beloved,
My dove, my perfect one,

For my head is wet with dew,
My hair with the drops of the night."

. . . My love thrust his hand
Through the hole in the door.
I trembled to the core of my being.
. . . Passion as relentless as Sheol.
The flash of it a flash of fire,
A flame of the Lord himself.

There is no sign of terror here, only a sacred glory of thrusting and trembling. And there is this recognition of women's erotic need from Exodus: "If he take him another wife; her food, her raiment, and her duty of marriage, shall he not diminish." The archaic King James phrasing can thwart contemporary understanding; the same line in more recent biblical language reads, "He must not neglect the rights of the first wife to food, clothing, and sexual intimacy."

From Paul in First Corinthians, in King James: "Let the husband render unto the wife due benevolence." Or, in a modern edition's version of "due benevolence": "The husband should fulfill his wife sexually."

A steady heat and urgency rises from the quills of the Bible's compilers in classical times and rises, too, from classical poetry and myth and medical texts. "Eros, again now, loosener of limbs, troubles me, uncontrollable creature," Sappho wrote. And Ovid's Tiresias, who lived as both male and female, declaimed that women take nine times more pleasure in sex. And Galen of Pergamum, physician to the Roman emperor and great anatomist of antiquity, pronounced that female orgasm was necessary for conception: a woman's climactic emission had to meet up with a man's. The contents of this female substance seem never to have been specified, but the requirement of ecstasy—a moment that appears to match our current definitions—was, for Galen, absolute.

For the next millennium and a half, until a few hundred years ago, Galen's understanding dominated science. A woman's "certain tremor" was a key to procreation for the fifth-century Byzantine physician Aetius of Amida. The Persian scholar Avicenna, whose eleventh-century *Canon of Medicine* was studied throughout the world, worried that a small penis might be an impediment to reproduction. The woman might not be "pleased by it," might not feel enough sensation to send her into blissful spasms, "whereupon she does not emit sperm, and when she does not emit sperm a child is not made." Gabriele Falloppio, discoverer of the Fallopian tubes in sixteenth-century Italy, stressed that a man's malformed foreskin might impede a woman's orgasm and impregnation.

How did Galen's thinking cling on so tenaciously? The longevity of his teaching is all the more baffling, given that only about one-third of women, nowadays, say they can climax through penetration alone. Were men and women of Galen's time, and long after, deftly attentive to the clitoris during intercourse? Better coached in the methods of vaginal orgasms? The shards offer up no answers. But, assuming that sexual skill was no better then than now, didn't women ever volunteer that they'd conceived without the tremor? Hints and theories of procreation without pleasure did emerge over the centuries, yet somehow Galen's wisdom wasn't supplanted. In the late sixteen hundreds, the widely used English midwifery manual titled *Aristotle's Masterpiece*, which asserted its scientific agreement with Tiresias about women's superior ecstasy, described the female role in conception this way: "By nature much delight accompanies the ejection of the seed, by the breaking forth of swelling spirit and the stiffness of nerves."

Still, this embrace of women's sexuality, from Exodus onward, shouldn't be taken as the prevailing ethos of any period. The ancient wariness and repression of female eros is a story that barely needs telling. There is Eve's position as first sinner: seductress and source of mankind's banishment from paradise. There is, from Tertullian, founding theologian of Christianity, the assignment of Eve's sinfulness to all women. All women were destined to be "the Devil's gateway." There are Moses's transcriptions of God's warnings in Leviticus. As the Jews encamp at Mount Sinai on their journey toward the land of milk and honey, God descends in a cloud and makes clear, again and again, that the center of a woman's sexual anatomy overflows with horror, with a monthly blood "fountain" so monstrous that she must be quarantined, "put apart for seven days, and whosoever toucheth her shall be unclean . . . and everything that she lieth upon shall be unclean, everything also that she sitteth upon." The litany of taint continues, relentlessly, until the decree that those who "uncover" the fountain and have sex will be expelled from the tribe, cast away from God's people.

For the Greeks, the original woman was Pandora. Molded by the gods out of clay, her erotic thrall and threat—her "beautiful evil . . . bedecked with all manner of finery" in the poet Hesiod's rendition, her "shameless mind and deceitful nature"—made her as dangerous as Eve. Lust-drunk witches of the Middle Ages left men "smooth," devoid of their genitals; and to the long line of living nightmares caused by female carnality, French and Dutch anatomists of the seventeenth century contributed the clitoris that grew with too much touching into a full-blown phallus, turning women into men who ravished their former sex.

But if the pre-Enlightenment West had always been frightened by female heat, sometimes extoling it, yet corralling it carefully within the bounds of marriage—where, for the sake of women's as well as men's sexual release, England's early Protestant clergy prescribed conjugal relations exactly three times per month, with a week off for menstruation—what followed eventually, with Victorianism, was a focused effort at extinguishing it. Lately historians have made the case that the Victorian era in Europe and America wasn't as prudish as we've tended to think; still, on the subject of female desire, it was a period of ardent denial. As with all the tectonic shifts of history, this one had uncountable reasons. One explanation has beginnings in the sixteen hundreds, with scientists' incipient realizations about the ovum, about the egg's part in reproduction. Slowly, incrementally, this ended Galen's legacy; gradually it separated women's ability to ignite from their ability to get pregnant. The ever-haunting female libido became less and less of a necessity. It could be purged without price.

Then, too, at the outset of the nineteenth century, nascent feminist campaigns and evangelical Christian rallying cries converged around the theme of irreproachable female morality. The two voices were intertwined; they amplified each other. Nineteenth-century feminists made humankind's salvation, here on earth and forever, their own womanly mission; Christianity made womanhood its exemplar. American prison reformer Eliza Farnham preached that "the purity of woman is the everlasting barrier against which the tides of man's sensual nature surge." Without this feminine barricade, "dire disorder will follow." And educational crusader Emma Willard proclaimed that it was for women to "orbit . . . around the Holy Centre of perfection" in order to keep men "in their proper course." One well-read American manual for young brides captured the inextricable feminist and evangelical spirits: women were "above human nature, raised to that of angels."

This was all a long way from "by nature much delight accompanies the ejection of seed." The innately pious had replaced the fundamentally carnal. The new rhetoric both instilled and reflected a transformation. In the mid–eighteen hundreds, in a letter about the sexual lapses of ministers throughout the Eastern states, Harriet Beecher Stowe wrote to her husband, "What terrible temptations lie in the way of your sex—till now I never realized it—for tho I did love you with an almost insane love before I married you I never knew yet or felt the pulsation which showed me that I could be tempted in that way—there never was a moment when I felt anything by which you could have drawn me astray—for I loved you as I now love God." And meanwhile, the renowned British gynecologist and medical writer William Acton was making plain that "the majority of women, happily for society, are not very much troubled by sexual feeling of any kind."

Yet beyond reproductive science, feminism, and religion, the Industrial Revolution had a tremendous impact on the West's thinking about what it meant to be female. Class barriers were breaking down; men could climb. This placed a value on work and professional ambition to a degree that may never have existed before, now that the rewards were potentially unlimited. And work—to borrow from Freud, who both was and wasn't a Victorian—required sublimation. Eros needed to be tamped down, libido redirected toward accomplishment. Victorianism assigned the tamping, the task of overall sexual restriction, primarily to women.

How far have we traveled in the last hundred or so years? In one way of seeing, Victorianism is a curio, encased in the past, its pinched rectitude easy to laugh at. This argument relies on a line of evidence leading rapidly away from the minimizing or denial of female sexuality, a line running through Freud's candid investigations of the erotic in women, through the brashness of the Jazz Age, the brazenness of flapper girls. It runs through the invention of the birth control pill, through the social upending brought by the sixties and the sexual revolution, and on through Madonna's aggressive cone-shaped breastplates and the pornographic self-displays of any number of lesser female celebrities. The opposing argument begins, too, with Freud, with the sections of his writing that render women as having, by nature, "a weaker sexual instinct," an inferior erotic capacity, and passes through post–World War I advice books like one informing that, unlike just about all males, "the number of women who are not satisfied with one mate is exceedingly small." From the forties and fifties, there is the story of Alfred Kinsey, whose research funds were revoked when, unforgivably, he turned from cataloguing the sex lives of men to publishing *Sexual Behavior in the Human Female*. Then, from the late sixties, there is the bestselling *Everything You Always Wanted to Know About Sex* delivering emotional law: "Before a woman can have sexual intercourse with a man she must have social intercourse with him." And finally there is the confluence between strains of contemporary thought: between the virginal edicts aimed mainly at girls and young women by evangelical Christianity, the waves of panic and sexual protectionism that overtake secular culture when it comes to girls but not boys, and the widely believed—and flimsily supported—thesis of evolutionary psychology that, relative to men, who are hardwired to hunt for the gratification of sex, women are rigged by their genes to seek the comfort of relationships.

This confluence is telling. In subtle yet essential ways, Victorian thinking about women and sex isn't so alien to our era. And science—evolutionary psychology—is an unlikely conservative influence. Mainstream evolutionary theory nimbly explains our physiological traits, from our opposable thumbs to our upright posture to the makeup of our immune systems. By contrast, evolutionary psychology, a field that has bloomed over the last few decades, sets out to use the same Darwinian principles to illuminate the characteristics of the human psyche, from our willingness to cooperate to our inclinations in one of the discipline's main areas of investigation, sex. The ambitions of the field are enticing and elusive, enticing because they hold out the promise that Darwin's grand logic can provide us with an all-encompassing understanding of ourselves, and elusive because the characteristics are so intricate and may have been created mostly by culture rather than inherited on our chromosomes. Evolutionary psychologists put absolute faith in the idea that our patterns of behavior and motivation and emotion are primarily the expressions of our genes. What *is*, evolutionary psychologists say, is meant to be, genetically speaking. This is equally true for the fact that we all have thumbs that help grasp and for the fact that—judging by appearances—men are the more lustful gender.

The role of social learning, of conditioning, isn't given much weight by the field's leaders. If promiscuity were considered normal in teenage girls and not in teenage boys, if it were lauded in girls and condemned as slutty and distasteful in boys, if young women instead of young men were encouraged to collect notches on their belts, how might the lives of females and males—how might the appearances that evolutionary psychology treats as immutable—be different? This kind of question doesn't much interest evolutionary psychologists like David Buss, a professor at the University of Texas at Austin and one of the field's premier sexual theorists. He dispenses with such challenges by amassing evidence that, all over the globe, male randiness and female modesty are celebrated. The widespread, in his view, proves the predetermined, the genetically encoded. Look, he has written in one of the discipline's academic manifestos, at the ideal number of sexual partners named by college students as they think forward over a lifetime; research has shown far higher figures for men than women. Look, around the world, at preferences in mates. From Zambia to towns of Arab Palestinians to America, societies set great value on chastity or some measure of propriety or reserve in women.

Evidence like this piles up in Buss's pages. And then he adds another worldwide mating reality—that from Zambia to America, financial prospects are prized in men—and this takes him to one of evolutionary psychology's pivotal conceits. Within the field, it is known as "parental investment theory." To the public, it may not be known by any name at all. And by most, the theory's components may be only hazily comprehended. Yet the conceit has traveled from academia through the media and into general consciousness. It has been fully embraced, deeply absorbed, become part of common wisdom. Parental investment theory goes like this: because men have limitless sperm while women have limited eggs, because men don't have to invest much of worth in reproduction while women invest not only their ova but their bodies, as they take on the tolls and risks of pregnancy and childbirth, because women then invest further in breast-feeding (the investment being in time, in extra calories required, and in the postponed ability to conceive another child)—because of this economy of input, far more pressingly relevant to our prehistoric ancestors, to our ever-endangered forebears, than to the humans of today, males have been programmed, since way back when, to ensure and expand their genetic legacy by spreading their cheap seed, while females have been scripted to maximize their investment by being choosy, by securing a male likely to have good genes and be a good long-term provider to her and her offspring.

This all fits neatly with the evidence from Zambia, Yugoslavia, Palestinian towns, Australia, America, Japan. And the theory's stark economic terms have a solid, incontrovertible sound. Our erotic beings, the differences in desire we observe between the genders, are the inevitable manifestations of evolutionary forces from eons ago. Parental investment theory gratifies one of our urgent longings: for simple answers about how we've come to be the way we are.

But the theory's foundation is precarious at best. Does the fact that women are expected to be the more demure gender in Lusaka and New York, in Kabul and Kandahar and Karachi and Kansas City, prove anything about our erotic hardwiring? Might the shared value placed on female modesty speak less to absolutes of biology than to the world's span of male-dominated cultures and historic suspicion and fear of female sexuality?

And then, what of Chivers's plethysmograph, which made a myth out of appearances? What of the drives that lie concealed beneath the surface, that crouch within the strictures? The sexual insights of evolutionary psychology can sometimes seem nothing but a conservative fable, conservative perhaps inadvertently but nevertheless preservationist in spirit, protective of a sexual status quo. Women, the fable teaches, are *naturally* the more restrained sex; this is the inborn norm; this is normal. And the normal always wields a self-confirming and self-perpetuating power. Because few people like to defy it, to stray from it.

One recent pop psychology mega-seller, *The Female Brain*, opens with lessons grounded in parental investment theory and serves as an emblem of the ways evolutionary psychology has spread its sexual vision throughout the culture. "The girl brain" is a "machine built for connection," for attachment. "That's what it drives a female to do from birth. This is the result of millennia of genetic and evolutionary hardwiring." The boy brain-machine is very different; it is built for "frenzies" of lust.

The book, like loads of others in the pop psychology genre, pretends to back its evolutionary theory with something concrete, with the technology known as functional magnetic resonance imaging, fMRI—with pictures of the brain at work. But the technology is nowhere near being up to the task. To spend time in fMRI laboratories, to stare alongside neuroscientists while fMRI data is sent from subjects' brains to lab computers, to listen as those neuroscientists strain to read and parse the pictures of brain regions forming on their monitors, as I have, and to ask bluntly about the state of our seemingly miraculous equipment, its capabilities much hyped by the media, is to understand that our technology is not at all precise enough to subdivide and apprehend the miniscule subregions and interlaced brain systems that enact our complex emotions, among them the wish to have sex. When, on the news or in a magazine, we hear or read something like, "The hippocampus lit up as subjects looked at photographs of . . ." we are learning something about as specific as a TV traffic reporter scanning from a helicopter and being able to say only, "The heavy traffic is somewhere in northern New Jersey." As scientists told me again and again, brain imaging just isn't a way to determine much of anything definitive about female versus male emotional neurology, not yet. And such technology may never be the right way to study *inborn* differences between the genders, because experience—use and disuse, positive and negative reinforcement—is forever altering neurological systems, strengthening some and weakening others.

Proclamations like the ones in *The Female Brain*—about connection versus frenzies, or about how a woman, to have satisfying sex, must be "comfortable, warm, and cozy" and, "most important," has "to trust who she is with"—are in striking parallel with the teachings of fundamentalist Christianity. The secular version is less extreme, but the messages are similar. As a pair of health education programs, designed by evangelicals and used in thousands of public schools within the past decade, instructed in their charts, the "five major needs of women" in marriage are topped by "affection" and "conversation." Sex is nowhere in the five. Across the page, the male list is led by "sexual fulfillment." In another graphic titled "Guys and Girls are Different," girls have an equals sign between "sex" and "personal relationship." Guys have the sign crossed out.

So, with scientific or God-given confidence, girls and women are told how they should feel.

CHAPTER FOUR

Monkeys and Rats

Her unruly red-blond hair tufting atop her head, Deidrah sat beside Oppenheimer. She lipped his ear. She mouthed his chest. She kissed his belly over and over, lips lingering with each kiss. After a while, he pulled himself up and strolled away from her attentions, glancing back over his shoulder to see if she was following. She was.

Deidrah, who was probably the most reserved female monkey in the compound, started in again on his white-haired torso as they sat together on a concrete curb. The habitat, a one-hundred-and-twenty-foot square, was filled with ladders and ropes and assorted apparatus donated by a local fire department and by McDonald's; an environment of trees and vines would have been too expensive to create and maintain. A trio of monkey children sprinted toward a tube, disappeared inside it, burst from the other end, and raced around for another run-through, berserk with joy.

From a platform on a steel tower, I watched with Kim Wallen, his beard silver, his eyes alight. A psychologist and neuroendocrinologist, he spent much of his time here at Yerkes, an Emory University research center outside Atlanta that was home to two thousand primates. We gazed down at the habitat's seventy-five rhesus, a monkey species that had been sent into orbit in spaceships, in the fifties and sixties, as stand-ins for humans to see if we could survive trips to the moon. Wallen had lived on a farm as a child when his father, a psychologist, decided to try out a utopian dream of cooperative goat-rearing. Wallen's observation of animal sexuality had begun then. He'd been watching monkeys now for decades.

"Females were passive. That was the theory in the middle seventies. That was the wisdom," he remembered the start of his career. Deidrah's face, always a bit redder than most, was luminous this morning, lit scarlet with lust as she lifted it from Oppenheimer's chest. "The prevailing model was that female hormones affected female pheromones—affected the female's smell, her attractivity to the male. The male initiated all sexual behavior." What science had managed to miss in the monkeys—what it had effectively erased—was female desire.

And it had missed more than that. In this breed used as our astronaut doubles, females are the bullies and murderers, the generals in brutal warfare, the governors. This had been noted in journal articles back in the thirties and forties, but thereafter it had gone mainly unrecognized, the articles buried and the behavior oddly unperceived. "It so flew in the face of prevailing ideas about the dominant role of males," Wallen said, "that it was just ignored."

What mostly male scientists had expected and likely wanted to see appeared to have blinded them. Wallen's career had been about pulling away the blinders. At the moment, below us, one female clawed fiercely at another, bit into a leg, whipped the weaker one back and forth like a weightless doll. Harrowing shrieks rose up. Four or five more monkeys joined in, attacking the one, who escaped somehow, sped away, was caught again. The shrieks grew more plaintive, more piercing, the attackers piling on, apparently for the kill, then desisting inexplicably. Assaults like this flared often; Wallen and his team usually couldn't glean the reasons. Full battle—one female-led family's attempt to overthrow another—was rare. That tended toward death: death from wounds and, some veterinarians thought, from sheer fright and shock. Occasionally the compound was littered with corpses.

When he thought about the way science had somehow kept itself oblivious to female monkey lust for so long, Wallen blamed not only preconceptions but the sex act itself. "When you look at the sexual interaction, it's easy to see what the male is doing; he's thrusting. It takes really focusing on the entire interaction to see all that the female is doing—and once you truly see it, you can never overlook it again."

Deidrah fingered Oppenheimer's belly, caressing, desperate to win his favors. He flopped down on his front, inert in a strip of sun. She kissed where she could get access, his ear again. The red of her face bordered on neon. She was near or in the midst of ovulation, her libidinous hormones high. When it comes to their cycles and sex, female monkeys are somewhere between lower mammals and humans; rhesus mating isn't limited to the time of ovulation, but in most situations, that's when it's a lot more likely to occur.

What was happening between Deidrah's ovaries and her brain as she stalked and stroked Oppenheimer is only partially understood, and the ways that biochemistry affects desire in women is even more complicated. Basically, though, sex hormones produced by the ovaries and adrenal glands—testosterone, estrogen—prime the primitive regions of the brain, territory lying not far from the brain stem and shared by species from *Homo sapiens* to lizards. This hormonal bathing then affects the intricate systems of neurotransmitters, like dopamine, that send signals within the brain, and this, in turn, alters perception and leads—in people and monkeys, in dogs and rats—to lust. The belief that animals, especially species less advanced than primates, don't experience lust, that their mating is scripted to the point of making them sexual automatons, is wrong, as Jim Pfaus, a neuroscientist at Concordia University in Montreal, would soon explain to me. Now, on the far side of the ladders and ropes, Deidrah was mouthing Oppenheimer's ear more and more ardently.

Bulky and torpid, Oppenheimer and the habitat's other adult male didn't fully take part in the life of the compound. They didn't belong to any particular family. They were merely breeders—and their peripheral status mimicked the male role in the wild. There, in Asian mountains or lowland forests, adult males lurked at the edges of female-run domains. The females invited them in to serve sexually. The males remained—desirable, dispensable—until the females lost interest in them. Then they were dismissed, replaced. In his compounds, Wallen removed the breeders and introduced new males about every three years, the time it took for them to become irrelevant, for their charms to wane, for the frequency of their copulations—almost always female-initiated—to fade. In the wild they seemed to stay attractive only slightly longer.

"Rhesus females are very xenophobic when it comes to other females," Wallen said. "Introduce a new female into the compound and she'll be hounded until she dies. But when it comes to males, females have a bias toward novelty."

With his pale muzzle and russet back, Oppenheimer loped off once more and Deidrah trailed him. A child of hers, less than a year old, hurried behind her. Wallen's assistants adored Deidrah. They loved her sprigs of out-of-control hair; they loved her personality, the quiet dignity she emanated most of the time, if not at the moment; and they loved the devotion of her mothering. Last year, upheaval in the compound had left her and her children vulnerable. Horribly frightened, they latched on to her and wouldn't let go. "Literally, she could barely get up and walk without being dragged down by her kids," Amy Henry, an assistant, said. "One held on to her tail. They wouldn't let her go. She accepted it all with grace. She knew it was her responsibility to reassure them that it was okay. She's always been a low-key monkey. But she gets very excited when she gives birth. And she gets very attached. I watched her carry her daughter on her back for a long time, right up to when she had a new baby. Not all moms will do that."

With hustling after Oppenheimer on her mind, though, maternal instinct was gone. She didn't seem to see let alone know her baby; she kept leaving it alone, and it kept having to scoot after her. She positioned herself in front of Oppenheimer, crouched, and tapped a hand on the ground in a staccato rhythm. She tapped like this persistently, the rhesus equivalent of unbuckling a man's belt. Yet her gesture contained a touch of hesitance. "She's being careful, because all the females around her are higher ranked," Wallen said. If they decided, for any reason, that they didn't want her having sex with him, they and their families might tear and bite her to death.

Wallen's realization, in the seventies, that rhesus females are the aggressors in sex had begun with a pattern he noticed in graduate school. At his university, pairs of adult monkeys—one female, one male—were observed in ten-by-eight-foot cages. At a lab in Britain whose work he read about, the cages were markedly smaller. On both sides of the ocean, the females had their ovaries removed; the scientists were tallying the copulations of the rhesus in the absence of ovarian hormones. And Wallen, who found himself contemplating the two sets of results against each other, was captivated by the fact that the couples in the tighter cages had a lot more sex. "So I pulled the literature on a range of similar tests that were done in different-sized cages, and the relationship was quite clear. In the smallest cages there was the most sex, in the largest ones there was the least, and the in-between ones had an in-between amount."

Soon Wallen arrived at Yerkes, and, watching the rhesus in the center's broad compounds—habitats whose size came closer to natural conditions—he developed his thinking about the way the tight confines in many experiments had helped to mold the accepted vision of monkey sexuality: lessening the female role and distorting the truth.

Put a male and female in a small cage, and no matter what the female's hormonal state—no matter whether she had ovaries at all—the pair would have plenty of sex, in part, Wallen came to understand, because their proximity to each other mirrored the kind of stalking Deidrah was doing now. This sexual signaling, created by the cramped dimensions, stirred the males to mount. The males *appeared* to be the initiators of the species. But put rhesus in a less artificial situation, and sex depended almost completely on the female's tracking, her ceaseless approaching, her lipping and stroking and belly-kissing and tap-tapping, her craving. Without her flood of ovulatory hormones, without the priming of her brain, copulation wasn't going to occur.

Are females the main sex-hunters in most other monkey species? The answer isn't yet known, Wallen said; not enough meticulous science has been done. Capuchins, tonkeans, pigtails—he named three types whose females are the stalkers. With their sweeping tails and ebony faces, female langurs initiate fervently. And among the massive orangutans, scenes like this were documented, for the first time, in the late eighties: males lying on their backs, showing off their erections to females, and waiting passively; and females closing in, mounting, pumping. As for bonobos, with their strangely parted hair and reputation for abandon, females avidly get sex going with males and with each other.

At last, with Deidrah tapping her crazed Morse code on the dirt, Oppenheimer reached out. Standing behind her, he set his hands on her hips. And suddenly she had what she sought, his swift thrusts. He pumped back and forth in a flurry.

Then he paused, pulled out briefly, touched her flanks, and slid inside her again for another bout of thrusting. He humped and pulled out repeatedly. When he came, thighs quivering and eyes going fuzzy, she twisted, turned her face to his, smacked her lips at high speed, reached back to seize him, and yanked him violently forward.

Her fulfillment was short-lived. Within minutes, she was hounding him again. At other moments, she might have moved on to the other male. "She has sex," Wallen said, about rhesus females on the whole, "and when he goes into his post-ejaculatory snooze, what does she do? She immediately gets up and goes off and finds another." Tracking the action of the compound, he asked himself, as he had so many times, whether the libido in women has similar drive, and whether "because of social conventions and imperatives, women frequently don't act on or even recognize the intensity of motivation that monkeys obey." He answered, "I feel confident that this is true."

Wallen didn't mean to imply perfect correspondence between Deidrah and the average human female. The distinctions included the impact of ovulation, so much more subtle in women. He and his former doctoral student Heather Rupp had been trying to grasp the ways that women's monthly hormones spur the neurotransmitters of desire. In one study, they had taken three groups of straight females and showed them hundreds of similar pornographic pictures—all featuring women with men—in three rounds, at different points in the women's cycles. Again, Wallen and Rupp used viewing time as a measure of the subjects' interest in the porn. One result was predictable: in the first round, the women who were near ovulation stared longer than the other subjects. But something else caught them by surprise. These same women, whose first round of porn came at mid-cycle, when testosterone and estrogen peaked, stayed riveted when they returned to the lab for their second and third rounds, as the month wore on and these hormones faded. The women whose initial viewing came during lower hormonal stretches didn't become transfixed when they ovulated. They continued to be less moved. Maybe, Wallen thought, some kind of conditioned arousal or indifference took hold. In later rounds, he guessed, the subjects still unconsciously linked the surroundings of the lab, the equipment, the porn to their reaction to their first viewing.

"One lesson," he said, "is that you don't want a woman to form her first impression of you when she's in the wrong menstrual phase. You'll never recover." He laughed.

Our conversation, on the platform above the compound, veered back to primatology, to the insights offered by our animal ancestors. He spoke about Deidrah's abundance of lust and about its constraint in women—about a communal sense of danger, a half-conscious fear of societal disintegration, that lay behind the constraining. And as I listened, and afterward as I dwelled on things, I thought of the historic terrors, the carnal archetypes: of witches, whose evil "comes from carnal lust, which in women is insatiable," according to the Christian doctrine of the Inquisition, "the mouth of the womb . . . never satisfied . . . wherefore for the sake of fulfilling their lusts they consort even with devils"; and of Eve, upon whose sinfulness all of Christianity is constructed, Eve, for whose evil the Son of God has to die, to sacrifice himself so that humanity can have a chance at redemption. This was the foundation, what lay beneath our culture's primary religion; it was imbedded in our societal psyche. And I thought, too, of monogamy: our inchoate idea that monogamy girds against social chaos and collapse, and our notion—the desperate inversion of our terror—that the female libido is limited and that women are monogamy's natural guardians. So we managed our fear.

Why, from beginnings in equally obscure academic publications, had parental investment theory come to permeate cultural assumptions over recent decades while monkey realities, ancestral facts, remained much less known? We embraced the science that soothed us, the science we wanted to hear.

"This organ serves a pleasure god," Jim Pfaus said. He held a plastic replica of the human brain in his hands. A Van Dyke beard and a hoop earring adorned his animated face. His expertise as a neuroscientist and his Concordia University labs were called on by the major pharmaceutical companies whenever they wanted to test, in rats, a new drug that might serve as an aphrodisiac in women—none had worked out in women so far. His labs sat in a university basement. There he studied his rats in a variety of cages and, in a surgical theater, removed their brains—about as big as a person's thumb from the middle knuckle to the tip.

Pfaus was obsessed with rat ways of seeing and feeling, learning and lusting, and when he wanted to investigate, say, exactly which set of neurons were sparked by a type of stimulation, by copulation-like prodding of the cervix or by the excitement of glimpsing a desirable male, one method was to provide a female rat with the experience, kill it, extract and freeze her brain, place the organ on a device resembling a miniature cold-cut slicer at a delicatessen, and shave off a specific, infinitesimally thin cross section. Peering at the slice through a microscope, he could pinpoint recent neural activity by noting the tiny black dots that told him where certain protein molecules—by-products of cell signaling—had been manufactured.

It was due to one woman that Pfaus—in his spare time the lead singer in a punk band called Mold—had been drawn to his specialty. Until the late seventies, scientists didn't study desire in rat females; they didn't see it; it didn't exist: as with the rhesus, scientists fixated on what the rat female did in the act of sex, not what she did to get there. And what she did in the act was go into paralysis. She froze in a position called lordosis, with her spine slung low and her butt cocked high, so the male could penetrate. Rat intercourse required female rigor mortis. It was easy to understand the female as absolutely passive, without will, a vessel whose involuntary perfume pulled the male in. Similar scientific ignorance had pervaded our idea of females across the animal kingdom, with "receptivity" being the key term.

But then Martha McClintock, like Wallen, helped take scientists deeper. McClintock had begun to make herself famous years earlier while still an undergraduate at the all-women Wellesley College. She built a case that women living in close proximity responded to each other's hormonal scents, causing the timing of their menstrual periods to converge, and her work was published in the revered journal *Nature*. Soon she was calling attention to female rat solicitations, to the female's specific hops and darts, her head-pointings and prancings away—her methods of inciting the male to put his forepaws on her flanks; to set his paws into the whir of flank-patting that instantly immobilized her, as though by hypnosis; and to slide himself into her. While Pfaus and I talked about this in front of a bank of his Plexiglas cages, one of his females went further, as regularly happened. Dealing with a stolid, sexually uninterested male, she stood behind him, mounted his backside, and humped, as though to put ideas in his head. How, Pfaus marveled, could science not have noticed this?

McClintock documented, too, that the female, if her cage allowed her to evade her partner, made sure to slip away from him, constantly, in the midst of his pumping, so the sex didn't end too quickly for her. Under any circumstances, in rat as in monkey sex, the animals attach, copulate, detach, and reattach repeatedly until the male ejaculates. The female rat, experiments showed, likes to prolong the process, to make it last longer than the male otherwise would. All of this, the solicitations and the preference for more drawn-out intercourse, suggested will and desire.

And McClintock established that by controlling the pace of mating, by getting the protracted stimulation and the rhythm that pleases her, the female can raise her odds of getting pregnant. She can raise them a lot. The extra thrusts, Pfaus said, cause contractions that aid sperm on their way into the uterus. And the *deeper* thrusts—for the male rat, forestalled from ejaculating, starts to pump harder—plies or jolts the cervix in a way that leads to a hormonal release that then helps to sustain a fertilized egg.

Pregnancy, though, is not an animal motivation, as McClintock and Pfaus, like Wallen with his monkeys, saw clearly. This was a critical point. Animal *species* have been designed by evolution to perpetuate themselves, to reproduce, but in the *individual* animal, it isn't reproduction that impels. The rat does not think, I want to have a baby. Such planning is beyond her. The drive is for immediate reward, for pleasure. And the gratification has to be powerful enough to outweigh the expenditure of energy and the fear of injury from competitors or predators that might come with claiming it. It has to outweigh the terror of getting killed while you are lost in getting laid. The gratification of sex has to be extremely high.

Pfaus had been following the early light shone by McClintock. Partly because of her research, he saw that a rat's brain was not merely a brain but a mind, that a rat's psychological experience could be a revealing version of our own. An array of experiments, of brain shavings, of injections of chemicals that boosted or blocked one neurotransmitter or another, of observations of rats making choices in all sorts of carefully constructed habitats and scenarios, fed his knowledge. One line of studies building on McClintock's work used a special cage with a Plexiglas divider down the middle. The divider had holes just big enough for a female rat—but not a male—to squeeze through. A female could determine the pace of sex by slipping from one side of the partition to the other and back again. "Female rats do what feels good. With the divider, she's having better sex. Better vaginal and clitoral stimulation. Better cervical stimulation." He described a study showing that intercourse stimulated the rat's clitoris: a colleague had painted males with ink, then charted the inky areas on their mates. About orgasms, Pfaus couldn't be sure whether female rats were having them; there was no easily measured sign, like ejaculation in males, to mark subjective explosion. But about pleasure and very intense desire, he was certain.

Proof ran like this: If, right after a rat finished a long-lasting session of mating, she was placed alone in another chamber, she would associate the new chamber with the sex she'd just had. Next, when given a choice between this new chamber and yet another, she would spend her time in the one linked with mating. She would make this choice even if the alternate chamber was set up to be much more inviting in other ways—even if the alternate space was dark, speaking to the nocturnal rat's sense of safety, while the chamber linked with pleasure was brightly lit, screaming of mortal danger. Run the same test with a female who'd just had quick—unsatisfying—intercourse and she would, afterward, opt for the dark space.

One of Pfaus's graduate students had lately performed and filmed a straightforward demonstration of desire—of motivation derived from the learned expectation of reward, just as desire develops in humans. Sitting with me in his office a few floors above his rat chambers, Pfaus played the video. The student picked up a female rat and, with a tiny brush, stroked the clitoris, which protruded from the genitalia like a little eraser head. She stroked a few times, then put the animal back down in her cage. Swiftly the creature poked her nose out of the open door. She clamped her teeth on the white sleeve of the student's lab coat and tugged the woman's hand inside the cage. The student brushed the rat's clitoris again, set her down again. And again the rodent bit into the sleeve, pulling, communicating unmistakably what she craved. This went on and on and on.

As we watched, Pfaus mentioned the anatomical oversights that had squelched our understanding of the clitoris—rat and human—until a decade before. The organ has sizeable extensions, lying internally in the shape of bulbs and wings. These are positioned, in part, just behind the front wall of the vagina. Yet these nerve-rich formations had gone mostly unnoted by modern anatomists, who either left them undrawn or gave them no import. Science seemed almost to have willfully diminished the organ, cutting it metaphorically away. It was another lesson in the minimizing of women's desire. Then, beginning in the late nineties, Helen O'Connell, an Australian urologist, detailed the organ's sprawl, its many inches in reach. And she championed its sensitivity to pressure through the vaginal sheath—sensitivity perhaps responsible for vaginal climaxes and possibly the explanation for the fabled and debated G-spot. O'Connell was blunt about the averted eyes of her scientific predecessors. "It boils down," she said, "to the idea that one sex is sexual and the other is reproductive."

Now Pfaus pulled apart his plastic model of a human brain, his fingers in the folds. He spoke about the neurotransmitters that define eros for women as well as men. The libido is, in a sense, two-tiered. There's the lower realm, in which hormones rise up from the ovaries and adrenal glands, float along the bloodstream to the brain, and fuel the production of the brain's neurotransmitters. How exactly this fueling happens is still a mystery; so is the quantity of fuel needed to keep the production line running well. The higher realm is the brain itself, the domain of the neurotransmitters. These biochemicals, not the lowly hormones, form the essence of lust.

Dopamine—its atoms arranged like an antennaed head with a spikey tail—is, in a way, the molecular embodiment of desire, its main chemical carrier. It isn't only that; it speeds through a multitude of the brain's subregions and exists in infinite relationships with other neurotransmitters and has all sorts of effects, from motor control (the trembling and sluggishness of Parkinson's patients stem from a shortage of dopamine) to memory. But dopamine is the substance of lust. And by way of his mini deli slicer, Pfaus had narrowed his sights on two tiny territories at the brain's primal core, the medial preoptic area and the ventral tegmental area. These were the heart of dopamine's sexual system, he said, "the ground zero of desire."

From this primitive epicenter, dopamine radiates outward. "A dopamine rush is a lust-pleasure," Pfaus continued. "It's a heightening of everything. It's smelling a lover up close—a woman inhaling that T-shirt. It's starting to screw; it's wanting to have; it's wanting more."

Yet for the excitement of dopamine to fix on an object, for it to be felt as desire rather than as a splintering into attentional chaos, it has to work in balance with other neurotransmitters. Serotonin plays an indispensable part. Unlike dopamine's keen drive, he said, serotonin dampens. Unlike dopamine's lust, serotonin instills satiation. Flood female rats with antidepressants—like the selective serotonin reuptake inhibitors, the SSRIs—that bolster serotonin, and the females will spend less time courting males. They will also bend their spines less fully, raising their butts less completely, to accommodate the males they do mate with.

It was important, Pfaus emphasized, to understand serotonin's virtues. They go beyond keeping depression at bay. The neurotransmitter also allows the brain's frontal lobe, more precisely the prefrontal cortex, the region of planning and self-control, to communicate effectively within the organ, to exert what's known as executive function. Serotonin reduces urgent need and impulse; it facilitates sensible thoughts and orderly actions. The problem, though, is that if serotonin is too strong in relation to dopamine, a woman making love is likely to find herself thinking about the next day's schedule rather than feeling overtaken by sensation and craving. But with serotonin and dopamine in the right balance, erotic energy will be neither displaced by tomorrow's to-do list nor permitted to fracture into chaos. With the frontal lobe and the libidinous core in harmony, desire can have both form and force.

For all the agility of his deli slicer, which could be set to the width of a micron, Pfaus was nowhere near to completely capturing the interplay of neurotransmitters. But a third type of transmitter essential to eros, he said, are the opioids, which surge with orgasm and also spike in tandem with dopamine's drive, so that glimpsing a lover's muscular chest or reading a paragraph of erotica offer a minor wave of opioid bliss. Describing this pleasure, he talked about the opioids' most potent varieties, the products of poppies: morphine, heroin. Send these drugs into the brain and satisfaction is so thorough—so much stronger than serotonin's well-being—that inertia takes hold. Both the executive region and the lustful center are quelled; direction and drive are nullified. In the less potent forms supplied by orgasm, the opioid rush lulls to lesser degrees. Meanwhile, a paradoxical process kicks in. Even while the opioids are quieting motivation, they are preparing the brain to be motivated again by somehow stoking the dopamine system. Orgasms simultaneously subdue the brain and teach it to seek more climaxes. Maybe even regardless of orgasm—for Pfaus couldn't be sure his female rats were climaxing—he saw the power of opioid bliss in his lab. Infuse female rats with a chemical that blocks this ecstasy, and they lose their desire to have sex at all.

Pfaus, whose hoop earring cast a gleam, whose mind bounced always between rodents and humans, translated this finding into a few words of advice: men had better perform; they'd better learn, they'd better deliver, and they'd better keep on delivering.

Not that this would solve men's problems, not that it should calm their worries. One morning he lectured his undergraduates on the Coolidge effect, a standard in sexuality textbooks, an expression of what Pfaus scorned as evolutionary psychology's "shtick." The Coolidge effect comes from a tale that goes like this: One day President Coolidge and his wife were visiting an experimental government farm. They took separate tours. When Mrs. Coolidge came to the chicken yard she noticed the rooster's frequent mating and asked the attendant how often this went on. "Dozens of times each day," he informed her, to which she replied, "Please tell that to the president when he comes by." The attendant did as she requested when the president arrived.

"Same hen every time?" the president asked. "Oh, no," the man answered, "different hen every time." And the president said, "Please tell that to Mrs. Coolidge."

The tale is used to hammer home the principle that male lust feeds on multiple partners. Pfaus mocked the faith that this is somehow less so for females. Rodent females, he informed his undergrads, do more hopping and darting to score with new mates. And they dip their spines deeper, so the new male has an easier time thrusting in.

During one of our talks, Pfaus swerved from the evidence he'd accumulated; he careened down a road of speculation. "When this generation of young people is fully studied," he said, words gathering speed, "we're going to see more supposedly male-like behavior, more women picking up men, more women getting laid and leaving, having sex without waiting to bond, more girls up in their rooms at their computers clicking on porn and masturbating before they get started on their homework."

It wasn't clear which age group he was thinking about, whether he meant girls who were now twelve or women who were now twenty-four, and it wasn't clear how he explained the unfettering he believed was underway, though it seemed to have partly to do with the Internet. Were there any concrete signs, I wondered, that the trend he imagined was real? Were girls and women staring more and more at the X-rated? Was their porn-surfing nearing that of men? There were only scattered answers, slivers of evidence. The most credible came from Nielsen, the consumer tracking company, in a report that one in three online porn users was female—four years earlier, the figure had been one in four. And porn-addiction counselors were quoted in the press saying that their ratios of female clients were rising. Yet the most vivid clue was probably James Deen's fan base.

Deen—who'd chosen his name and its spelling himself—was a porn star who'd shot two thousand scenes over the last eight years, scenes in which a delivery man is asked inside for a blow job, in which a principal teaches a lesson to a new high school teacher, in which a chained and gagged blonde submits or a MILF has her way, scenes made, like most of the films produced by the thirteen-billion-dollar-a-year porn industry, with men in mind. But the scenes had caught on among teenage girls. Teens and young women seemed to make up most of his tens of thousands of Twitter followers. They watched him on PornHub and Brazzers and Kink.com; they traded his images, set their computers to pick up any mention of his name, sent him proposals of marriage. A profile on *Nightline*—"the young man your teenage girl may be secretly watching," the introduction intoned, "a porn star for the Facebook generation"—added to the craze. *GQ*, the *New York Observer*, the *Guardian* in England swarmed next. Some fans said they were attracted by his boy-next-door looks, some by the way he held a woman's eyes amid doing everything else, but along with his slender build and the possibility that he gave slightly more eye contact than the average porn stud, the basics were the basics: a maximally sized erection, a minimum of dialogue, a dose of violence ("I've been into rough sex pretty much my whole life," he told one interviewer, "so I'm not, like, bad at it"), lots of female moaning, many genital close-ups.

Like Deen's popularity, Suki Dunham's nascent success as an entrepreneur pointed to changes taking place. In her case, the changes ranged over age groups. In a New Hampshire town of four thousand, in a farmhouse bordered by a white picket fence, with a tree house out back for her two kids, Dunham designed state-of-the-art vibrators like the Freestyle and the Club Vibe 2.OH.

The predecessors to her devices had been around for over a century—initially as aids to doctors and nurses who believed they needed to massage patients to "paroxysm" as a cure for hysteria—but during the last few decades the percentage of women saying they've used a vibrator has gone from one to over fifty, and a few years ago vibrators appeared on the shelves at Walmart, at CVS, at Duane Reade. Trojan spotted an opportunity, entered the market, advertised its Tri-Phoria on TV, and watched sales leap in a period of cataclysmic economic decline. Durex, another condom maker, did the same and had the same results. And Dunham, who grew up in a coal-mining town in Pennsylvania, the daughter of a man who ran a small excavation outfit, was developing a high-end niche.

Dunham, who has an eager, thin-lipped smile, had worked at Apple for nine years, marketing iMacs. One Christmas in her mid-thirties, she received in her fireplace stocking an iPod and a vibrator from her husband, who traveled a great deal on business. Using the two devices together started her thinking, and now she had a warehouse packed with merchandise. She was shipping to thirty countries and had just begun teaming with a reality TV star who was turning the products and Dunham herself into a regular part of the show. Dunham's company sold sleekly curved vibrators, aggressively bifurcated vibrators, discreetly streamlined vibrators, and an app that let the user program a vibrator to pulse and tremble to the rhythm of an iPod's music. The newest item, the bullet-shaped 2.OH, slipped into a bullet-shaped pocket stitched into a thong. The 2.OH massaged, at an array of intensities, to the music of a club or party. It had enough charge to last three hours.

With females hopping and darting in front of us, Pfaus asked, "Why do we have Pandora's box—why have we boxed in women's sexuality? Why do we keep women's desire relatively repressed? We men are afraid that if we open the box, open her control, we're opening ourselves to being cuckolded. We're afraid of what's inside."

He laughed over a memory of mine: not too long ago, porn on cable TV in New York had, by law, included a blue dot. Always, the dot covered the male member. Women's bodies were thoroughly exposed, yet the dot floated wherever the penis went.

And I thought, as well, of trips I'd made in other parts of my writing life, trips to places where the blue dot took other, not so comical forms. Once, in a remote village in northern Kenya, I'd asked a group of Samburu men why their culture practiced clitoridectomy. They answered matter-of-factly, "So our wives will be faithful."

Sometime later I spoke again with Chivers. She'd been designing a new experiment that would use subliminal images as another means to tunnel beneath consciousness and perhaps beyond culture. She outlined the study, then said that she'd been thinking about the scene in her college lecture hall twenty years before, about the sound of "Eeew," about that fleeting, unforgettable, half-funny syllable as an expression of a history and prehistory of "all kinds of prohibitions and restrictive perspectives on female sexuality." Suddenly her voice leapt: "Look at all the barriers! Look at all the obstacles! But that isn't what amazes me."

She was a fastidious scientist who chose to surround herself with surfaces of barely adorned cinder block, to spend her days in an almost monk-like cell, to avoid pronouncements, to let her data speak for itself. Now, though, she ignored the scientific restraints. "Those barriers are a testament to the power of the drive itself. It's a pretty incredible testament. Because the drive must be so strong to override all of that."

CHAPTER FIVE

Narcissism

One wall of Marta Meana's cramped university office was covered with postcard-sized reproductions, portraits from past centuries. All were of women. The faces of Vermeer's *Girl with a Pearl Earring* and *Portrait of a Young Woman* floated on their dark backgrounds, their skin luminous, their eyes turned toward something or someone behind them.

Below them, as a joke, Meana handed me a picture of two control panels. One symbolized the workings of male desire; the other, female. One had only a simple on-off switch; the other had countless knobs. "Trying to figure out what women want," she said, "is a real dilemma." It was a dilemma she was trying to address—as a scientist, as a couples counselor, and as the president of the Society for Sex Therapy and Research, the most prominent organization in its clinical field—in different ways than Chivers.

And it was a dilemma that Isabel, a lawyer at a nonprofit,

was attempting to solve for herself as she debated whether she should stay with her boyfriend of eighteen months, Eric, and marry him if he proposed, which she sensed he would. The issue was that despite his good looks, his intelligence, his kindness, and his skill in bed, she rarely wanted to make love with him.

Isabel, who was not one of Meana's patients but was one of the women I spent my time learning from, recounted a scene from the evening of last year's Valentine's Day. In her small Manhattan apartment, Eric had run her a bath, strewn it with salts, surrounded it with candles, and considerately left her to lie alone in it. When she stepped out of the bathroom, she found her bedroom lit with more candles. And on her bed was a deluge of rose petals arranged in the shape of a heart. Through a fair amount of conscious effort, she managed to be seduced by this gift: while Eric took her place in the bathroom and showered, she lay back on the bed, brushed the petals across her lips, and dropped some onto her shoulders and breasts. When he emerged from the shower, she did feel pleasure as he went down on her, slid his broad shoulders up her body, and slid inside her. But the pleasure was precarious. And on many other nights what she felt was merely patience or something worse than that.

She loved him, she was certain. She said, "I remember the first time I brought Eric home to St. Louis to meet my dad and my stepmother and my grandmother. She's eighty-eight years old. She helped raise me when my parents were divorced. We call her The Patter. If she sits next to you, she's going to find some part of you and pat it. She's going to pat your hand or your knee or your arm. Pat-pat-pat as you sit there. She's incredibly special to me. She's incredibly affectionate. She's deaf, mostly. I think that's one reason she's so tactile; her ability to communicate is hampered. There's something very child-like about her. And one afternoon during that visit, I walked into the living room, and she and Eric were sitting on the couch, holding hands. He looked totally comfortable. There was nothing ironic in his expression. I think they had been talking, but talking is labor-intensive with Granny, and now they were watching TV. Probably she'd been patting him, and they'd wound up holding hands. I think most men would have been highly self-conscious. Holding hands with her like that would have been an ironic exercise. But for Eric it was natural."

She also said, "The shades in my bedroom let in a small amount of light, and he likes to sleep with something over his face. A T-shirt, a pillow, an arm, all three—I don't know how he breathes. It's kind of hilarious. In the mornings, I have to peel away layers to get to where his face is. I want eye contact." She endured sex once a week but yearned for this every day. "I'll find his eyes under everything and wait for his eyelids to open, and then I'll find space for my body right against his."

She was transfixed by the tenderness of his gaze and tormented by the fact that her lust for him had waned within a few months of their starting to date; she sensed that his proposing might happen any day. She dreaded it. She was in her early thirties. She believed that she couldn't afford to make the wrong choice, and amid all the truths she tried to weigh rationally, it was impossible for her not to compare her situation with Eric to the two years she had spent with her previous boyfriend. When she had dressed for Michael, she had selected her clothes while subjecting herself to a silent inquisition. "Am I a doll?" she had asked herself as she stood in front of her mirror, or as she lingered in a boutique's fitting room, deciding whether to buy. "Am I a fantasy?" Michael's preferences weren't extreme, but they were pointed. High-heeled boots, a short skirt. Or tight jeans and a T-shirt slung halfway off one shoulder, sizeable hoop earrings, dark eyeliner.

He was ten years older. And he was particular, though his requests were never demands. What she put on was her choice. He let her know, keenly, precisely, what he liked: the black lace bra through which her nipples showed. But the decision about whether to fulfill his wishes belonged entirely to her.

The problem was that she wanted to fulfill them all, though his taste in clothes was not hers. What was she collapsing into? she had berated herself. Yet it didn't feel like collapse. There was strength in sliding on the lace thong that matched the bra, in pulling on the jeans or skirt, the boots. He would be riveted. She had that power. An alertness spread through her body as she dressed for him. An awareness suffused her skin.

With Eric, she didn't have to accuse herself of any capitulation. He liked what she liked, and she counted this as a sign. When they went out in summer, she often wore a loose-fitting pastel green dress she'd bought on a trip to Guatemala. It was girlish, she knew, and she laughed at herself because of it. But Eric cherished this quality in her. To be who Michael had wanted required stepping off a precipice, dismissing the voice that warned her against inhabiting his wishes, plummeting over that edge. Women who dressed with urgent, ungoverned need for the desire of men could set off, inside her, a flurry of disdain, like an instinctive aversion to a weakness or wound. Yet whenever she walked into a restaurant where Michael waited for her at the bar, his focus seemed to pluck her from the air, midfall, and pull her forward. His eyes held a thoroughly different kind of constancy than Eric's later would. Eric adored her. Michael admired her. She was a possession, the heels of the boots she picked for him taking her across crowded rooms toward her owner. The boots were like the frames and pedestals he chose for the photography and sculpture in his gallery. He had specific opinions about how she was best displayed.

Her mind was already reeling by the time they sat down to dinner, yet she kept the appearance of balance. The display that pleased him depended on a degree of agility. In conversation and body she maintained dexterity, but when his breath or hand grazed across her in any way, or even when there was no contact at all, only proximity, she could become so frantic with need she grew almost angry. "If you don't touch me right now, I'm going to scream," she would plead silently. "Please, God, touch me right now. Please, God, something's got to be done here."

She came quickly, repeatedly, when the dinners were at last over and they were in bed. The certainty of her coming guaranteed it; she didn't have to doubt, so doubt never got in the way. Her mind never obstructed; it had been unspooling since the evening's start.

Michael's effect on her had been all the more enthralling because of the way she viewed her own body. At the age of seven, anointed a flower girl for a summer wedding, she had worn a dress of pink flounce and lace trim with a pink sash and a crown of roses and baby's breath. She couldn't have been more pleased; the outfit was the prettiest she had ever put on. But when she glimpsed the girl chosen to walk alongside her, who wore the same flounce and trim and sash, and who seemed half Isabel's size, the spell cast by the fairy-tale clothes floated away, replaced by bewilderment, then despair, that two seven-year-olds in identical gowns could look so far from identical. Ever since, she'd seen herself as encased by a soft excess, sometimes horrifically thick, sometimes subtle. She assaulted this flesh by dieting, or ignored yet never forgot it. And though as an adult she told herself that it had been years since anyone could rightly have called her even chunky, still there was this padding that dismayed her. Under Michael's stare, she had felt pared away. The sharpness of his eyes had somehow cut her body to a better outline. Eric didn't have that ability. He was gentle, while Michael had been gentlemanly; he was empathetic, while Michael was at once solicitous and commanding. Michael's admiration had convinced her of her allure; Eric's telling her she was beautiful couldn't quite make it so.

The relationship with Michael had ended only because she understood he would never commit to her, never marry her or even live with her, but it didn't end cleanly at all. Months after breaking things off, she met him for dinner, and afterward, outside, he turned up the collar of her overcoat, hailed her a cab, and five minutes later sent her a text: I'm following you. Soon she was buzzing him into her building. There were lapses like that, lapses and lapses. The end point she had announced to her friends took forever, it seemed, to become fact, until she could no longer bear to confess her failures to them.

"I could not part with him," she said. "I couldn't get him out of my skull."

Meana was a psychology professor at the University of Nevada at Las Vegas, and just before I flew out to meet her, she said that we should start by going together to a Cirque du Soleil show at one of the casinos. So, soon after my plane landed, we sat in a darkened U-shaped theater and began our conversation while a pair of topless, dark-haired women in G-strings dove backward into a giant water-filled champagne glass on stage. The women plunged in from opposite sides of the pool, swam toward each other, and entangled with each other, eel-like. They slid up the walls, arching their spines and dragging their breasts along the glass.

Next, a wispy blonde came skipping like a schoolgirl out on stage. Wearing a tiny pleated skirt, she swirled her hips and kept a set of hula hoops spinning around her waist. Suddenly a cable snatched her up above the audience, hoisting her high. It was her act's climactic moment, a symbolic ravishing. The nymphet opened her legs wide above our eyes, splitting them wider than seemed humanly possible; the splitting was almost violent.

Then a sinewy black woman wearing only beads thrust and pumped her gleaming body to a tribal beat. The soft-porn performances followed each other in fast succession, the stage dominated by arresting women. The audience was divided equally between the sexes. Finally the platinum-wigged MC cried out, "Where's the beef?" and a long-haired man in a cowboy's vest and chaps climbed through a trap door. He strutted and swiveled and bared his abdominal ladder of muscle. He shed the chaps, kept only his groin covered, and stood in his cowboy boots, flexing his ass. Yet even as male nudity had its minutes, a dozen female bodies surrounded him.

In her early fifties, Meana, wearing a shirtdress and tights that evening and wearing her bronze-colored hair in bangs, didn't doubt the usual explanations for the fact that women far outnumbered men among the performers, though she didn't believe they were terribly illuminating. Those explanations went like this: The men in the audience would have been made too uneasy by more male nudity on stage. For them—for the heterosexuals among them, anyway—the cowboy needed to be obscured by breasts. And for the women in the crowd, the female nakedness fed an addiction—judging their own looks against iconic beauty. So the ticket buyers were gratified, given a live version of what they were used to from a million images on billboards, in magazines, on television: for the men, an opportunity to lust; for the women, a chance to compare.

Meana saw more in the imbalance on stage. She began simply, with something that fit with what Chivers had found through her plethysmograph, as the flaccid Adonis tossed stones on the beach. "The female body looks the same whether aroused or not. The male without an erection," Meana said, "is *announcing* a lack of arousal. The female body always holds the promise, the suggestion, of sex." The suggestion sent a charge through both genders.

And then, the imbalance served women in a further way, an essential way—a way that formed the crux of Meana's thinking. To be desired was at the heart of women's desiring. Narcissism, she stressed—and she used the word not in damning judgment but in plain description—was at the core of women's sexual psyches. The females in the audience gazed, erotically excited, at the women on stage, imagining that their own bodies were as searingly wanted as those in front of them.

From her office wall, in one of the two Vermeers Meana had put up, the young woman glanced back and outward, her thin-lipped mouth smiling timorously, as if she couldn't be sure anyone was aware of her. In the other, the more full, parted lips were not smiling at all. The girl had no doubt she was being watched.

"Being desired is the orgasm," Meana said somewhat metaphorically; it was at once the thing craved and the spark of craving. Her confidence about this narcissistic engine arose partly from a curtained area in her lab, from a contraption that seemed to belong in an ophthalmologist's office. With chin clamped and immobilized on its little cradle, her subjects peered at a screen, at a series of soft-porn images. The contraption took hundreds of readings per second of how the eyes roamed and where they paused.

For a few years, she had been comparing female and male patterns of focus. Long ago, she had earned a master's in literature and planned on a career teaching great novels. But she found she couldn't endure standing in front of a classroom and trying to make her students share what she felt. "I didn't want to tarnish it," she said. So she had returned to study as an undergraduate, to get the background she needed to start on her doctorate in psychology.

In a study she had just published, her heterosexual subjects had gazed at pictures of men and women in foreplay, among them a couple standing at a kitchen sink, he behind her, tight against her, genitals out of sight, the two of them wearing little more than a few suds of dish soap. Viewing the sequence of pictures, the male subjects stared far more at the women on the screen, at their faces and bodies, than at the men. The females looked equally at the two genders, their eyes drawn to the faces of the men and to the women's bodies—to the expressions, it seemed, of desire in the men and to the flesh desired in the women. For the females, heat seemed to radiate from the men's urgency and from the women's power to generate it.

Meana wanted to test this further. She had to be certain that the females weren't checking out the women's bodies merely for the sake of comparison with their own. She had to dispense with this reason that she saw as intertwined but secondary; she had to confirm that her subjects were staring at what turned them on.

One method might have been to run the same experiment while her subjects masturbated. That way, their eyes would more surely seek out what carried an erotic jolt. But there was little chance that she could get this kind of study approved by the university's review board and, even if she could, too high a chance that research with masturbating women would bring Meana the outraged scorn of Las Vegas's conservative press—condemnation that could, in turn, endanger all of her explorations. Las Vegas was a paradoxical place, with nearly all its advertising based on titillation and with prostitutes waiting at licensed ranches down the highway and yet with a prudish strain in the atmosphere, a resistance to the animal impulses that made people flock to the city. This divided psyche seemed an extreme version of a split that ran throughout the country. It left the erotic as hard to study as it was omnipresent. This was partially why Chivers, who had earned her PhD and done the first of her plethysmographic experiments in the States, had returned home to Canada after graduate school to continue. During her years in the United States, her research had been the target of ridicule. The *Washington Times* had protested the American government dollars that helped to pay for her work. "Federally Funded Study Measures Porn Arousal," the headline had read. A congressman had demanded an investigation. The outcry over her small project soon faded, but she worried about an acute American aversion to looking too closely, too carefully, at the sexual.

Meana was going slowly with the review board. She was designing a study that would use eye-tracking and X-rated videos. The films would turn on her subjects more than the soft-porn photos could. In this more heightened state—a state less conducive to cognition and comparison—would the women's eyes become less drawn to female body parts and more compelled by everything male? She didn't think so. She expected that the pattern from her earlier experiment would hold, that when it came to bodies, the female figure would prove full of electricity.

As she developed the new study and hoped for the review board's approval, she didn't yet know about Chivers's research with the pictures of isolated genitals. Those results might have led her to wonder if, in her video experiment, women would seek—as well as female bodies—Deen-like erections, pure declarations of male desire.

Meana's ideas grew not only from her lab but also from her work as a clinician, some of it trying to help women besieged with dyspareunia, genital pain during intercourse. The condition is not, in itself, caused by an absence of lust, yet her patients reported less pain if their desire increased. So part of her challenge was how to enhance desire, and despite prevailing wisdom, the answer, she said, had "little to do with building better relationships," with fostering communication between patients and their partners.

She rolled her eyes at such notions. She described a patient whose tender lover asked often during sex, "'Is this okay?' Which was very unarousing to her. It was loving, but"— Meana winced at the misconception behind his delicate efforts—"there was no oomph," nothing fierce, no sign from the man that his hunger for her was beyond control.

Talking with Meana made me think of Freud and one of his followers, Melanie Klein. Sexologists don't have much time for psychoanalytic theory; they tend to ignore or deride Freudian ideas as ungrounded in the empirical research that defines their discipline. Meana never mentioned Freud, yet his thinking, and Klein's, seemed to float within hers. It seemed to hover, as well, behind Chivers's readings of blood.

For Freud, sexuality was etched into our psyches with our earliest rapture—and the mother's breast was the dazzling source. "It was the child's first and most vital activity," he wrote a century ago, "his sucking at his mother's breast. No one who has seen a baby sinking back satiated from the breast and falling asleep with flushed cheeks and a blissful smile can escape the reflection that this picture persists as a prototype of the expression of sexual satisfaction in later life." The primal need for sustenance dictated the child's first lessons in eros; survival and sensuality converged. "The child's lips, in our view, behave like an erotogenic zone, and no doubt stimulation by the warm flow of milk is the cause of the pleasurable sensation." The infant's consciousness was flooded, immersed in moments of nearly orgasmic power.

"The finding of an object is in fact a refinding of it." Freud delineated the way our adolescent and adult desires took shape. We searched for the past, for the pleasures we once received, which were given by the mother not only in feeding but—all the more so by her, as opposed to the father, during Freud's lifetime—in countless other ways of tending to the baby, from cleaning its genitals to nuzzling its neck to clutching it tight. "A mother would probably be horrified," Freud continued, "if she were made aware" that she was "rousing sexual instinct and preparing for its later intensity. She regards what she does as asexual, pure . . . after all, she carefully avoids applying more excitations to the child's genitals than are unavoidable in nursery care." She should, Freud assured, "spare herself any self-reproaches even after her enlightenment. She is only fulfilling her task in teaching the child to love."

The erotic energy of girls, in Freudian theory, was soon led along intricate emotional routes and rechanneled from mothers to fathers. But the original lessons lingered; the mother's sexual pull was never erased.

Then Klein amplified Freud's thought. For Freud, nursing and the breast were far less important, finally, than the phallus or lack of it, in writing the psychic scripts of boys and girls. Klein toppled this hierarchy. The breast, for her, was nothing less than mountainous. Maybe it was inevitable that Freud, as a man, would magnify the phallus above all and that it would take a female psychoanalyst to overturn this. Maybe Klein's thinking arose not only from her being a woman but from working, as a clinician, with young children, from observing the psyche close to its beginnings, rather than reconstructing childhood through the lives of grown patients, as Freud did. No matter what the reasons, Klein evoked a breast that seemed to occupy the infant's entire vision. All else disappeared. The breast soothed and withheld, seduced and denied, gave itself and guarded itself, taught love and rage. It was "devouring . . . bountiful . . . inexhaustible . . . persecuting"—it consumed our earliest consciousness and never really relinquished its overwhelming role.

Freud believed that homosexual attractions churned within women because of their experiences as infants; his and Klein's writing offered an explanation for the pulsings of blood when Chivers's women watched women together, women alone.

The breast was the first locus of desire; a woman's body was its owner; we are all on a quest of "refinding."

And the mother in Freud and yet more in Klein added depth to Meana's thinking about women's sexual narcissism. Through the female bodies in her lab or on the casino stage, through a mostly nude model washing dishes at a kitchen sink or topless swimmers diving into a giant champagne glass, women made themselves, unconsciously, vicariously, recipients of the unmanageable desire they themselves had once felt for the bodies of their mothers. They acquired their mothers' erotic omnipotence.

On one of the laboratory walls, outside the curtained room where women's pupils were tracked, there was a poster from an Annie Lennox concert Meana had been to. Lennox's piercing, incantatory voice, her unflinching lyrics, her band's icy, electronic sound, seemed almost audible sometimes as Meana spoke. "Sweet dreams are made of this; who am I to disagree," Lennox sang. She then laid out, without judgment, without lament, some of the inescapable realities of lust. Meana's face was round while Lennox's was lean; Meana's bangs were pixieish while Lennox's hair was shorn half an inch from her skull; Meana's voice didn't hold the singer's unremitting insistence. But there was a shared impatience with the tales people tell themselves about desire. Meana's features were nimble, expressive; her mouth twisted occasionally, faintly, into something akin to a grimace. This happened when she talked about the legion of couples counselors who held to the idea that, especially for women, incubating intimacy would lead to better sex.

Empathy, closeness—these were supposed to be the paths. For Meana, these paths might lead to lovely places. Lust, though, wasn't likely to be one of them.

"Female desire," she echoed Chivers's experiment with the strangers, the close friends, "is not governed by the relational factors that, we like to think, rule women's sexuality in contrast to men's." She was about to publish a study built on long interviews with women whose marriages were sexually bereft. It might be right, she said to me, that bad relationships can kill desire, but good ones don't at all guarantee it. "We kiss. We hug. I tell him, I don't know what it is," she quoted from one subject. "We have a great relationship. It's just that one area"—the area of their bed.

It was important to distinguish, Meana went on, between what was prized in life and what was most potent as a source of lust. Women might set a high value on ideals of togetherness and understanding, constancy and permanence, but "it's wrong to think that because relationships are what women choose, they're the primary source of women's desire." Again, she spoke of narcissism and the wish to be the object of primal need.

The attainment of this wish, she argued, required not closeness but a measure of distance. An object of lust was, by necessity, apart. She warned against the expectation or even the hope of reaching popular romantic dreams: of "merging" with a partner, of being able to say "you complete me." This was the wrong standard for love. This kind of bond, or just the striving for it, could suffocate eros. Melding left no separation to span, no distance for a lover's drive to cross, no end point where the full force of that drive could be felt.

"Sometimes we wake up looking at each other," Isabel said. There was a radiating warmth in this perfectly timed stirring, this simultaneous opening of eyes with pupils and irises so close they were about to blur, she and Eric on the verge of vanishing in proximity. Second best and still wonderful was the lifting of coverings from his face so his eyelids opened and she was seen, recognized, taken in, ensconced, absorbed.

Why, she asked herself, indicted herself, interrogated herself, did she feel indifference—why, if she was honest with herself, had she begun to recoil when he reached out in a way that asked for more? It made no sense to her. At the party where they had first met, she had been the one to spot him first; on their first date, she had been the one to kiss him first; during their first months together, she had, she said, felt such lust she had "climbed him like a tree." Now, at a year and a half, she "clung to him like Velcro," had the daily thrill of his just-woken eyes, and felt as though her desire had been stolen, spirited away by some mischievous minor god.

She took action. She ventured into an upscale sex-toy shop and bought massage oil, a blindfold. This was with the intention not of blocking out his handsome features but of transforming the effect of his touch. Attempts like these were successful, slightly, temporarily. What was wrong with her? Sometimes, she said, she wished he would "take more of the marauder approach"—her shoulders pinned to bed or wall, her nipples bitten hard, her thong pulled harshly aside, torn. But she told herself not to ask for this. "Because he would feel badly and because his gestures would be empty, a parody of what I want. The whole thing is that it should be instinctual. The idea that I would have to request it . . ." Her voice trailed off. Was it possible, she asked herself, to have both what she'd had with Michael, for whom the marauder approach had been one part of a hypnotizing repertoire, and what she had from Eric, the profound sincerity, the absolute presence? What was she setting herself up for if she stayed with him? Did she need to extricate herself, no matter how excruciating that would be?

Early in her second winter with him, a great snowstorm hit New York. It piled high plumes on the railings and layered plush cloaks on the sills. It pushed drivers off the streets and consumed their cars once they were parked. The blizzard caused a communal thrill, all the more so since only a few days remained till Christmas. Several days before this, she and Eric had put up her tree, adjusting and clamping it in the stand and adorning it. As she had hooked a gleaming red ball to a high branch, her eyes teared abruptly with gratitude that she was doing the decorating with him.

And now, in the middle of the Saturday afternoon blizzard, she came home from shopping for gifts and, in her kitchen, talked with him about what she'd bought. She noticed that he wasn't saying much, then that he wasn't taking part at all. He walked out of the room, into the vestibule.

He stopped, turned. She realized his hands were behind his back. Maybe, she thought, he'd got her an early present. He stepped again into the kitchen and knelt on one knee.

"What are you doing?"

"I'm asking you to marry me."

"You're doing that? Right now?"

Plainly, he was, because below her, in his outstretched fingers, he held a ring. Still, she seized on the thought that he might be joking, because the knee was so sudden and the kitchen, as a setting, was so strange.

"Are you going to speak?" he asked.

She didn't.

"Are you saying yes?"

So much hope lay in that question, and it was met by her own, hers full of desperation to preserve everything she had with him.

"Yes," she said, "I am saying yes."

She joined him on the kitchen floor. She slid on the ring, a diamond in a deco setting, a hexagon. He'd chosen it without any hint from her. As ever, they had the same taste. He told her that, hours earlier, he'd called each of her parents and asked for their blessings. She loved that, too.

On the linoleum they hugged and drank the bottle of champagne he had ready. He listed all the reasons he wanted to spend his life with her, and eventually they stood and moved from the kitchen. They went not to the bedroom but outside, into the evening dark, into the unabating snow. As they walked, the sills grew more and more lush, the cars more and more enveloped. Everything was covered over, buried.

CHAPTER SIX

The Alley

For her twenty-fourth birthday, Ndulu gathered with several friends at a restaurant downtown. The restaurant was a straightforward, undramatic place, and Ndulu lived a straightforward, undramatic life, but a few of her friends were gay men who did not worry as much as she did about what was and wasn't appropriate. In addition, the dinner involved some drinking.

So, near the end of the meal, David called their waiter over and informed him that Ndulu needed a birthday kiss. By the time David was halfway through this overture, she had ducked her head and was covering her face with her hands. David had no idea how perfectly the waiter's looks matched a longing of hers. And neither David nor Ndulu could have known how his wishes fit her fantasies.

Standing now behind her, he didn't laugh at David's request. He didn't tell David no, and he didn't give Ndulu the suggested kiss. Instead, he leaned down over her shoulder with his lips close to her ear. "Go into the bathroom," he said softly, though not so quietly that her friends didn't glean his words.

She stayed in her chair. Her friends—especially David, an aspiring musician who was used to plenty of sexual conquests and who campaigned regularly against Ndulu's reserve—were jubilant, riotous with this development. They pushed at her with their hands. They pushed at her with their words. They managed to send her in the direction of the wooden bathroom door.

These are fantasies, the first harbored by Isabel, the rest by others:

"My grammar school principal. I'm in a skirt. Eleven or twelve years old. He has silver hair; he's overweight; he's wearing a blazer. He finds a way to call me into his office. He's married. He has a million reasons not to do this. It's not that perversely I think he's attractive; it's that I'm attracted to the fact that he's so attracted to me. He's risking that someone might walk into this office; he's risking his job to be with me."

"A shower in a hotel with multiple people."

"A random guy on the street. I don't want candlelight."

"Oral sex with a man I can trust. I know that sounds mundane, but I suppose this stems from growing up in conservative, backwater, buttfuck Kentucky, where blow jobs were expected and relished in discussion but eating out was either gross or wasn't discussed at all."

"I am a young virgin peasant girl whose family is one of many that works the land of a rich landowner; the landowner or his son forces himself on me, and I know I have no choice but to let him do what he wants. Or I am the school whore or a social misfit, and the football team is taking turns with me. I am still coming to terms with the fact that things that I find to be wrong—rape; taking advantage of those without power—are the things that bring me to mind-blowing orgasms."

"Not scenes. Textural sensations playing through my head."

"Another couple having sex, near me, where I can see them. Someone licking me or touching me, maybe two people, and then a man entering me from behind. I wouldn't say it's violent. Maybe vigorous—is that a dorky word to use?"

"The rape scene from *The Accused*, I'm ashamed to say."

"A married, older man that I work with, who I'm not even all that attracted to, fucking me from behind against a whiteboard—we work at a school—and hitting my face against it. Then he turns me around so that I can fellate him. Him coming on my face."

"Once in a while I dream of dreamy stuff: kissing and fluffy desserts we feed to each other. Quite often I dream of many men, servicing me all at once."

"A stranger, usually a construction type, peeking in my window."

"Essentially rapes. I started masturbating when I was around ten or eleven—the most common one back then was a middle-aged bald man while I was chemically paralyzed. Receiving pleasure wasn't my fault if I was being raped; I didn't have to explain myself to Jesus or my parents. Where the bald man came from I have no idea. Then, when I started having sex with my husband, it turned out that orgasm was kind of a lot of work. It was very important to him that I have one whenever we had sex, and sex with him was nice, but orgasm required that I fantasize. Reenter the bald man."

"Thinking about the girl-on-girl ads on Craigslist."

"The bored housewife who lets the FedEx man take advantage of her, only to be seen by the postal delivery guy, who forces himself on her next. The bored teenager who pretends to fall asleep while lounging by the pool in a loosely tied string bikini while a construction crew just happens to be working close by."

"The part in *Excalibur* where Arthur's father transmutes into the form of another man and has sex with Arthur's mother while wearing bloody armor."

"I used to have rape fantasies, but now they have been replaced with walking into a room and seeing the man I'm dating sitting on a chair with a college-age girl straddling him, facing him. She's always thin with big breasts—so stereotypical, I'm sorry—and long, shiny hair all the way down her back. He has one hand wrapped up in that hair and the other fingering her anus."

"Earlier in my life it was wooing: parks and lots of looking at the moon. The violent aspect did not develop until later when the malaise of my first marriage settled in. I remarried and here's the deal: I'm super competent. I run the house, do ninety-nine percent of the kid care, have a PhD and a successful career. I am absolutely in control *all the time*. In bed, fantasy allows me to feel out of control while being in control. I don't give myself away, but I imagine giving myself away. 'A sordid boon,' Wordsworth would say. No, I imagine myself *taken* away. I would like my life to have more of that: I'd like my husband to *take* control. But he's not able to. I don't know if it's the no-means-no message he's gotten since health class in middle school. So I create a world in my head."

"Being tied up and blindfolded while someone I love shares me with a number of people I can't see. Multiple people desiring me and concentrating on me alone. Or if I'm feeling particularly tired or unhappy and my body just isn't responding, I'll make it rougher. This releases me from all the other thoughts, from worrying if my son did his homework and when the mortgage is due."

"A man eleven years younger than me, a boy really, who I had an affair with. I'm married ten years this week. I'm thirty-eight. We only saw each other maybe once a month, sometimes longer between. And we never talked by phone or email unless we were arranging to meet. Now I've broken it off. And I buy all kinds of outfits for him that he'll never see. The way he would look at me when I opened the door is what I hunger for. Or the afternoon he taught me how to really give a blow job, in my backyard next to my pool, the sun shining on us—I have never in my life wanted a cock in my mouth as badly as with that man. In replays now that it's over, I slide my mouth over a dildo I keep hidden. With my husband, I'm just making love."

"What I'd like to do with my boyfriend. A public place, a subway platform, a park."

"The noise my lover makes when she climaxes."

"We're a conservative couple and my husband is the only man I've ever been with, so when I close my eyes, his body is the one I have an image of. Lesbian sex, adulterous sex, I'll find myself wandering sometimes into the forbidden, but I always go back. His body is simply erotic. It's mine. I know it. I understand it. I have fantasies that I whisper to him in bed about tying his hands behind him and making him watch me masturbate. I always think it's funny that people who find out I was a virgin—by choice!—when I got married think I'm naïve or prudish. If only they were in my head."

"Males and females, males more when I was single, females more now. Images as mild as the curve of a hip or as hard-core as full-blown bondage."

"Older brother–younger sister incest (I should add that I'm an only child)."

"A visit to a male gynecologist, with me naked in the stirrup-things. The doctor inserts various instruments; he fingers me to make sure there's nothing wrong with my cervix. A sexy female nurse starts examining my breasts. Young male medical students come in to watch, to be taught how to conduct the pelvic exam. The doctor instructs the nurse to play with my breasts, to make sure that arousal is functioning normally. He checks my clit. I start squirming with pleasure. I'm vulnerable and completely exposed to a figure of knowledge and authority. Or being raped. It's a twisted paradox, but in my head rape equals control equals trust. I don't have to worry about anything, because the other person has power over me, I know he could kill me, so it's his responsibility to make sure that I'm safe. The rapist is often a soldier, Serbian or Russian, not American, because of the stereotypes about Eastern European men being dominant and rough. He's always a stranger. He uses his own strength, as opposed to a rope or gun, to control me, usually by pinning my wrists above my head against the floor. At first

I don't want it, and struggle against him, but he knows when I start to enjoy it. Occasionally I fantasize about being raped as punishment for having anti-feminist fantasies."

"An older man sitting on a chair and masturbating while I have sex."

"I've always battled with my weight. So being someone else and looking completely different than I actually do. Sex with a celebrity, sex with a cute bartender from the other night, sex on stage, with one spotlight and one chair like in *Cabaret*. The feeling that I am desire in the audience's loins."

"The first fantasies I can recall involved having sex with men in their twenties or thirties. I had found some porn magazines of my father's. I was around eleven. My favorite scene was of a man in his thirties approaching me from behind and pushing me up against a chain-link fence, pushing my clothes aside but always having a firm grasp on my body. Now my fiancé is in Iraq. Ninety-five percent of my fantasies involve him. We have the photos we send each other. I hear I'm kind of a small-celebrity army pinup."

"My boss; a stranger in a bar; my father's friend. Horny and demanding and forceful. So consumed by me that he can't help himself. . . . As an undergraduate I felt like I had to monitor my internal and external life toward consistency. In other words, if I truly believed in women's equality with men, then I'd have to have sex and imagine sex that reflected that—no domination, no rape fantasies. One result was that I married a nice liberal man who shared my convictions on how sex should be. Seven years later we divorced."

"A really sexy girl lies back on my bed. I grind against her face with my vagina, making her eat me out kind of violently."

"Rape—which, until very recently, I had trouble admitting even to myself. It seemed to fly in the face of all my participation in Take Back the Night rallies in college, all those women's studies courses. Men take turns holding me down."

The appeal of rape—in the mind, in the lab—haunted Meana and Chivers and took our conversations to uneasy places. Two of their sexologist colleagues, Jenny Bivona and Joseph Critelli at the University of North Texas, had gathered data from nine earlier studies and offered a sense of how commonly women turn themselves on in this way. "For the purposes of the present review," Bivona and Critelli spelled out, "the term 'rape fantasy' will follow legal definitions of rape and sexual assault. This term will refer to women's fantasies that involve the use of physical force, threat of force, or incapacitation through, for example, sleep or intoxication, to coerce a woman into sexual activity against her will." Depending on the study, between around 30 and 60 percent of women acknowledged that they took pleasure in this kind of imagining. The true numbers, the authors argued, were probably higher. The subjects conjured the scenes while they had sex, welcomed them while they masturbated, daydreamed about them.

One explanation invoked the same reasoning as the woman who said, "I didn't have to explain myself to Jesus." Rape fantasies removed guilt. Women embraced them to escape the shame imposed, from the beginnings of girlhood, on their sexuality, to escape the constraints imposed going back and back in time.

Another theory took imagining and relishing rape as a type of taboo-breaking.

An experiment carried out at an amusement park by Cindy Meston, a University of Texas at Austin psychology professor, contributed to yet another explanation. Hundreds of heterosexual roller-coaster riders were shown photos of the opposite sex; the subjects were asked to score, in Meston's words, "dating desirability" before and after the ride. The thrill of fear spilled over into eros: following the ride, the scores rose. The phenomenon, which Meston labeled "excitation transfer," hinted at interweaving circuits of terror and sexual arousal within the brain, and perhaps made sense of what one woman told me, that she felt as though her rape fantasies had an immediate physical effect, that they coursed straight to her groin, causing the contractions of orgasm.

There was anatomical logic to the idea that calling up thoughts of rape and feelings of fear—or feelings of shame brought on by transgressing taboo—could quickly provoke the spasms of climax. The theory belonged to Paul Fedoroff, a psychiatrist at the University of Ottawa's Institute of Mental Health Research, who treated paraphilics, people whose main erotic compulsions fall far outside the norm: fetishists, exhibitionists, zoophiles, sexual serial killers, pedophiles. Like so much surrounding our under-researched sexual selves, Fedoroff's reasoning was backed by informed speculation rather than proof, yet his theory had resonance. Some of his patients, he had told me, when I was researching a book about paraphilias, seemed to suffer from what he called a "sticky switch" governing their parasympathetic and sympathetic nervous systems. These are two branches of our autonomic circuitry, the wiring that regulates our automatic functions, like heart rate, sweating, and salivation. The parasympathetic system controls arousal while the sympathetic sends us into climax. "The natural progression during sex," he said, "is that the parasympathetics are set off, and at some point when we become sufficiently aroused a switch flips and the sympathetics kick in and we start to have an orgasm. But the poor paraphilic has a sticky, sluggish switch and needs to do something extreme to get the sympathetics going." Besides orgasm, the sympathetic system takes over in situations of emergency. Fedoroff's idea was that some paraphilics use the deviant, the forbidden, to stoke their sense of danger or mortification—to create an emotional emergency, put extra pressure on the resistant switch, open up the sympathetic paths, and propel the brain and body into ecstasy.

Many of Fedoroff's patients were convicted criminals, but he told me about a case that wasn't criminal at all. A heterosexual couple had sought him out; the woman could no longer climax, not with her partner. She'd taken to having sex with a series of men in the same night, to watching videos of women having sex with animals, to making videos of herself masturbating—these sent her toward orgasm. Climactic sex with her partner seemed a hopeless cause "until," Fedoroff wrote in a journal article, "it was discovered that she consumed large amounts of L-tryptophan, available in health food stores, to help her sleep. This substance is metabolized into serotonin, which is known to cause difficulty reaching orgasm. She was advised to discontinue taking L-tryptophan. Soon afterward, her ability to reach orgasm through intercourse with her partner returned, and with it, her paraphilic interest in group sex, exhibitionism, and zoophilia disappeared."

According to Fedoroff's theory, fantasies of sexual assault might well serve, for some non-paraphilic women, as a way to unstick the switch; they might supply an emotional emergency and enable orgasm.

But for Meana, rape fantasies were rooted in the narcissism that was imbedded in the female sex drive. As we talked, she narrowed her ideas into an emblematic scene: a woman pinned and ravished against an alley wall. Here, in her vision, was an ultimate symbol of female lust. The ravager, overcome by craving for this particular woman, cannot restrain himself; he tears through all codes, through all laws and conventions, to seize her, and she—feeling herself to be the unique object of his unendurable need—is overcome herself.

Right away, she regretted what she'd described, the alley image she'd called symbolic. She hadn't used the word "rape," but the scene evoked it.

"I hate the term 'rape fantasies,'" she said quickly. The phrase was paradoxical, she stressed; it had no meaning. "In fantasy we control the stimuli. In rape we have no control." The two ideas couldn't coexist.

"They're really fantasies of submission," she continued. She elaborated on the pleasure of being wanted so much that the aggressor is willing to overpower, to take. "But 'aggression,' 'dominance,'" she sifted through the terms that came to her as she tried to express the wish. "I have to find better words.

'Submission' isn't even a good word." It didn't reflect what women were imagining as Meana's scene culminated: their willing acquiescence.

Yet she looked vaguely stricken; she knew her parsing of language couldn't tame the subject. The fantasy of the alley, no matter how much she focused on vocabulary, retained its aura of violence. And as Bivona and Critelli had pointed out, the paradox in logic, in conjuring one's own lack of control, didn't exactly mean that the fantasizing woman wasn't immersing herself in an experience of sexual assault. The assault wasn't real, of course; the immersion was only partial; but the violence, the overpowering were being lived, if only in the mind. The fantasies occupied a realm that was both infinitely far from the actual and yet psychologically close to it. Were they different from any of our other intensely felt and yet idle wishes? To commit crimes and become rich? To inflict awful harm on our enemies? We don't act on these imaginings, and in one sense we don't want to transpose them from the world of the mind to the world of the real. We don't want that at all; we would then have entered a nightmare. Yet our fantasies do speak of our desires.

When Meana first talked with me about the alley, I was interviewing her for a magazine article. Just before the story went to print, we spoke on the phone. She suggested a change in the way I presented things: I should specify that it wasn't a stranger who pressed the woman against the wall; it was someone she knew.

I didn't remember this detail from our discussions. I asked if she was sure the change would represent the truth of her thinking. She hesitated. She worried that without the addition, the scene would seem very much a rape and that she would appear to be endorsing this kind of attack. I assured her that I had made the difference plain: between the gratifications of the invented and the horrors of the real. But she was in agony. She believed that all of her work would be reduced to the alley, that it was all people would remember. That dark place seemed to loom in her mind, almost as if it really was the only thing she had spoken about. We went back and forth about noting that the man was not a stranger. I asked her, Who exactly was he?

And we discussed possibilities: his being a date; his being someone the woman had just met. But there was no nailing it down. It seemed more faithful, not only to her thinking but to the variations within women's fantasies, to leave the man undescribed except for the force of his desire.

We agreed not to make the alteration, she in deep discomfort, still wishing somehow to soften the scene. When the article came out, she was barraged. Her in-box filled with hundreds of emails. Oprah asked her to be on her show. "I've become the overpowerment lady," Meana said to me later, when I met with her again in Las Vegas. The alley wall had been central in the reaction to her words. And some of the reaction was vehement. "There was hatred. People said I was part of the machinery that puts women down, that I was inciting men to rape."

Yet there were plenty of other responses. Oprah, as she introduced Meana, voiced her own troubled feelings about the alley but played, at the start of the segment, a taped interview with a perky, middle-of-the-road woman who echoed the allure of Meana's scenario. And the emails were also full of gratitude.

"There were lots of messages from high-powered women thanking me for allowing a discussion of elements of sexuality that don't fall neatly into an ideological box," Meana said, relieved. "One woman, in the art world in New York, told me, 'I could not say what you said without feeling shamed, as though my eroticism made me a willing participant in a patriarchal system.'"

Still, Meana remained unsettled. All the attention had churned up something submerged, a latent distaste about studying sex at all, a shame about it, a fear of it. "Even we who do this research have internalized the culture's sexophobia. It was fine when it was just me in my lab, me with my students. But with the spotlight on—no. Suddenly I was asking myself, Why was I studying something so inconsequential? Why couldn't I be studying depression? Why couldn't I be studying suicide? I had to stop myself. I had to remind myself, In what way is sex inconsequential?"

She paused. "I have no insecurity about my feminism," she said. "I feel on solid ground. What I said in the article stepped outside what have become the conventional, comfortable ways of talking about female sexuality, the soft ways, the ways that leave everyone feeling good, not anxious. I don't think what I said was misogynistic. I don't think it was harmful. Now, do I know whether certain turn-ons excite only because of a social structure that disempowers women? Whether certain fantasies are an eroticization of disempowerment? No, I don't know. But I do see the world from a feminist perspective. And part of that is wanting women to be able to be who they are sexually."

She sounded almost at ease. She seemed almost to have

located solid ground. Yet the footing seemed unsure, as if at any instant it could turn treacherous. The alley was no place to stand.

Did the fantasies, as Meana asked, "excite only because of social structure"? What about the narcissistic longing that lay beneath, that led to the grammar school principal, to the landowner's son, to fantasizing about the rape against the pinball machine in *The Accused*—was this "an eroticization of disempowerment"? She raised the quandary that was always near: culture or genes?

To think back to Deidrah was to see an immense societal impact. How else but culture to explain the vast difference between Deidrah's aggressive sexuality, her stalking of mates, and women's desire to be desired, which dictated the pleasure of being chased? Men made objects of girls and women; girls and women, living in a male-run world, absorbed the male outlook as their own and made objects of themselves. Hadn't culture taken Deidrah's drive and, in women, both partially quelled and completely recast it?

Yet when Meana contemplated the psyche, she called herself an essentialist, mostly. About the interplay between nature and nurture, she placed more weight on the inborn. She placed the weight gingerly. Her essentialism was a hunch, a sensibility; there was no way, she knew, to measure the inherent against the acquired, not for the time being; there was no way to assign a percentage to its role in narcissism, in rape fantasies. (A wealth of pop psychology writing declares confidently that there is an all-determining link between inborn levels of testosterone and myriad forms of aggression or passivity—sexual forms high among them—in men and women. Genetic factors give boys and men a lot more of the hormone, as counted in the bloodstream, and this makes boys and men a lot more aggressive. But among the list of problems with this seductively simple logic is evidence that comes, again, from Deidrah. Compared with male rhesus, females have as little testosterone as women do in contrast with men. Yet female rhesus run the sexual show, incite warfare, and rule the world of rhesus politics.)

Meana's intuitive leaning toward the innate added to her uneasiness about the appeal of the alley scene. Emphasizing the genetic meant that there was no escape; it meant that the allure was fundamental.

Chivers was haunted in a similar way. She saw the culture's relentless sculpting of women's sexuality, but her mission was always to look past that, to seek and examine what lay beyond society's reach, and this put her into a wrenching confrontation with rape. She knew about emerging results from a close colleague's experiment: genital blood flow spiked when women listened to rape scenes in a lab. (An experiment of her own demonstrated, as well, that situations of fear or euphoric excitement triggered no vaginal pulsing if sex wasn't involved. In one comparison, she played videos of a woman being chased up a flight of stairs by a rapist or by a rabid dog. Only the sexual scene flooded the genitals.) She dwelled on studies of victims that documented not only lubrication but sometimes orgasm during sexual assault. And she remembered—from her postdoctoral program in Toronto, when she had done work as a therapist—rape survivors who'd confided their own arousal, their own climaxes, to her.

How to understand this? How to comprehend this harrowing evidence? Was something deeply scripted, something intrinsic, at work?

Chivers felt that it was. And she helped to develop a reassuring theory: that prehistoric women had been constantly subject to sexual attack, and that the ability to lubricate automatically in reaction to all sorts of sexual cues evolved genetically as a protection against vaginal tearing, against infection, against the infertility or death that might follow. Genital arousal might not represent desire, she argued, but might, rather, be part of a purely reflexive, erotically neutral system, a system that was somehow intertwined with but separate from the wiring of women's libidos. And the instances of orgasm might reflect nothing more than friction.

Yet the theory of separate systems was elaborate, precarious. It defied more straightforward thinking: that being wet meant being turned on, that there wasn't much that was neutral about it, just as was true for men and being hard. Gradually Chivers settled on what had perhaps, she told me, been obvious all along, that it was possible to be stirred by all sorts of things one didn't, in fact, want. By sex featuring bonobos, by sex featuring assault.

"I walk a fine line, politically and personally, talking frankly" about rape, she said. "I would never, never want to deliver the message to anyone that they have the right to take away a woman's autonomy over her own body. Arousal is not consent."

* * *

This was one of Ndulu's fantasies: "A faceless white man slams me against a wall and holds me in place with his elbow as he strokes his rock-hard dick. He whispers into my ear all the vile things he wants to do to my body. He tells me he's going to shove his cock so far into my pussy I'll feel it in my belly; he says if I don't behave, he'll call in his friend, who's right outside, ear pressed to the door, violently masturbating, to come fuck me as well. Would I like that? he asks. Would I like two hot cocks in me? He takes me rough and hard from behind, standing up. Just when he begins to call out loudly as he comes inside me, his friend bursts in and comes on my ass. Both men are calling out in such pleasure that it almost sounds like they're crying."

This was the way Ndulu's imaginings usually went, and the violence of the men, the unrestrainable lust of the men, the ecstasy of the men that poured out in their "almost crying" were made more heated for her—and terribly painful for her—by race. Ndulu had grown up on American oil company compounds in West Africa and Europe, gone to college in the American Midwest, and now lived in New York, where she worked as a graphic designer. Over the course of her childhood, her adolescence, and her young adulthood, she had learned to believe that her skin and hair and features added up to an overall appearance that fell somewhere between tolerable and not. This was true, above all, about the shade of her complexion. "In winter," she said, "it's medium. In summer, though, no matter what I do, it gets dark. In summer, I can't even look at myself."

She spoke of how her mother had always made it clear that lighter skin was more attractive than darker. During her own childhood, Ndulu's mother had watched *her* mother's adoring eyes on the paler face of Ndulu's aunt. "In black families, there's always this issue," Ndulu said. "It's no different in Africa. My aunt was the belle of her village, because she was so light. My grandmother spent all her time on her."

As a teenager, Ndulu had done what all the girls of her West African city did, what she had begun to learn from her mother before she could talk. Into her hair, to make it less kinky, she slathered a grease that was the pale yellow of custard. "It wasn't as thick as butter, but it was thick, and it was oilier than butter, and you had to put a lot on. It would drip down the sides of your face in the sun."

These days, in New York, she was trying to wean herself from oil by wearing her hair fairly short. But she hadn't yet quit; she didn't expect to. "It's so common. I don't think I know a single black woman who doesn't use it. It's just something we have to do. To make our hair look more white. I hate it. It reminds me of what I am and what I'm not."

She added, "I've read *The Bluest Eye*," and she talked about the lessons of Toni Morrison's novel. "I know how I should be, I know the way it's supposed to go—the whole empowerment thing. In college I wrote essays out the wazoo about everyone being equal and equally beautiful. I don't feel any of that."

Her college was close to 100 percent white; her friends were an insular group of black women. They talked often of the black pop stars and students they fantasized about, of the superiority of black men—the size of their penises, the hairlessness of their skin. Her gay friends now—white, Asian—did the same. And all the while what she felt was that to be the focus of a white man's violent need—"all of my fantasies are of a white man; and except when he is faceless, he is beautiful, beautiful beyond words; he is tall, with azure eyes and thick, dark hair"—would be to know, in the most absolute way, that she was desirable.

The waiter was tall enough, broad-shouldered, with blue eyes and dark hair. "He was gorgeous," she said later. She stepped into the bathroom; he followed, turning on the faucet, opening the tap fully, the water loud. How much noise is kissing going to make? she asked silently, as they began. He leaned back against the wall, pulling her toward him. She braced her palms against the tile on either side of his shoulders; his fingers spanned her ass. At some point, he slid his cock out of his pants; she felt it rigid near her waist. She wished she was the one with her back to the wall, but it didn't matter—the thought was crushed by the strength of his hands.

The faucet went on providing its white noise. "Suck it," he told her.

Even more than his features, his voice now seemed to spring directly from her imagination, her moments of private lust: the way the two words he repeated held not even the undertone of a question.

She lifted her hands away from the wall, straightened, took a step back. Again, he told her what he wanted.

"I have to go," she said.

"No, you don't."

"I have to go."

"Stay."

When she tried, when she turned, she couldn't get the lock unbolted.

"I've been drinking," she protested. "I have a boyfriend."

"Do you really?" He held her forcefully.

"I have a boyfriend," she lied. "I need to go."

Something shifted in his face, and when he spoke again the presumption was thoroughly washed from his voice, as though by the impact of a wave. He looked disoriented, lost. "Okay," he said. This time, she managed the lock.

Her friends were in a clamor when she emerged. They assumed she'd gone further than kissing. David insisted on a description of his cock. He often regaled her with the dimensions of his conquests. "I'm not going to talk about that," she said. Seconds later, she confessed that she hadn't carried it through. When they groaned, she apologized, and when they asked why, she answered that she didn't know. "I just couldn't," she told them. Then she went home and lay down and let the scene unfold—differently, from the moment of his demand, from her inability to unbolt the door—as she touched herself, let it unfold until she came, let it splinter her mind, obliterate her, obliterate her again the next morning, again the next night, again on more mornings and nights than she could count.

CHAPTER SEVEN

Monogamy

Alison's husband, Thomas, was a youth league basketball coach. He taught the pick-and-roll, the defensive stance, the proper way to catch a pass, the correct preparation for a free throw. He believed in fundamentals. He believed that if his eleven-year-old players learned nothing else, and if they never touched a basketball again after their season with him, their practices and games under his tutelage would be worthwhile if they gained a set of twelve basic basketball skills, or at least recognized their importance. Life, in his opinion, was a matter of fundamentals as well, and his hope was to have some part in getting kids ready not to win at a sport but to thrive in the years ahead. He was also a corporate lawyer. But he looked forward to the Blazers' Wednesday evening practices and Saturday morning games a good deal more than to anything he did at his high-paying job.

Alison knew the twelve skills by heart, or anyway nine of them, or at any rate she had been able to list nine four years ago when their son, Derek, had begun his basketball career. But two years ago, Derek had retired. He had become the official G.M., scorekeeper, trainer, and unofficial towel boy for his father's team, and since then Alison's recall for the fundamentals had dwindled.

Derek's retirement had been brought on by his realization, as a fourth grader, that he just wasn't much of a player, that he was not only shorter and chubbier but slower and less coordinated than his teammates. He raised this in a matter-of-fact way with his parents. When he said that he would prefer a spot in "the front office," they laughed, talked the decision through with him, hugged him, and agreed. Yet in his first season in his new position, Alison had gradually stopped coming to practices and soon also to games, because of her own work as an attorney, she told her husband and son, and because Derek's younger sister was getting old enough to have her own schedule. What Alison suspected, though, what she suspected with something awfully like certainty, was that she wasn't merely avoiding the sight of her son draping towels around the shoulders of the boys and girls on the team (the league was coed) or the cooing, complimentary remarks she received about Derek from the other moms, but that she was avoiding a new—no, a partially new—vision of Thomas. She just didn't want to see him teaching the box-out technique or charting another play on his clipboard in a time-out huddle.

Then, as Derek's second front-office season was about to start, her son begged her to watch the opener. So, after Thomas had cooked Saturday pancakes, scrubbed the griddle, loaded the dishwasher, and driven off with Derek to the New Jersey community center to make sure all was ready for the Blazers' arrival, Alison helped Derek's sister pick out a special outfit and followed in her car.

A circle and a line defined a debate in sexology, a debate about the natural course and velocity of female desire, a dispute entangled with a question: how well do marriage and monogamy work for women's libidos?

Rosemary Basson, a physician and professor of psychiatry and gynecology at the University of British Columbia, had started devising and drawing the circle over a decade ago, sketching it for female patients and couples, for women worried about their lack of lust. She was just past sixty now, feathery brown hair cropped above her ears. Her voice was wispy, her skin pale. As we talked across a coffee table in her Vancouver office, she wore a flowing skirt with a pattern of leaves; she seemed almost formless, ethereal. Yet there was something quietly stern, no-nonsense, governess-like in her speech. She'd been pulled toward the field of eros as an internist in England. Assigned to a ward of patients with spinal-cord injuries, a floor with a steady supply of men left paralyzed by motorcycle accidents, she found herself confronting, now and then, a man who had worked up the courage to ask how—or whether—he could ever have sex. She asked a supervisor for advice. "Change the subject," he told her. "Change the subject." She still remembered his tone: tight, panicked. She'd been dealing with the subject of sex ever since.

Pen in her white fingers, she drew her circular flow chart for me. Proudly she recalled its first journal publication. Now the psychiatric profession's bible, the *Diagnostic and Statistical Manual of Mental Disorders*, the *DSM*, was about to enshrine her thinking in its pages. A massive volume filled with criteria for conditions from autism to sexual dysfunction, the *DSM* distinguishes the normal from the abnormal. In her diagram, Basson rendered a picture of women's desire as intrinsically slow to build. It was the result of a series of decisions; it was hardly a drive at all. "We're just not talking about an innate hunger," she said.

The intricate diagram was meant to evoke the step-by-step progress of a successful sexual encounter for women, beginning with a box at the top of the circle. Inside the box—the outset of the encounter—was the phrase "reasons for sex." Raw desire wasn't likely to be the reason, though the chart allowed for it as a possibility. Much more probably, Basson said as she sketched, the woman was going to make a deliberate calculation based on the hope of outcomes like "feeling positive emotionally, feeling loved." About two-thirds of the way around the circle, the words "arousal" and then, finally, "desire" appeared; at this belated stage, physical sensations, pleasure, and wanting took over, to some degree. But this depended, she explained, on the partner showing "respect," on the woman feeling "safe," on the couple's being in "an appropriate context," on the partner's touches being considerate, being just right. Listening, it was hard not to imagine flowers given, a bedroom with the lights low or off, a wife with basically cuddly inclinations, a husband's gentle caresses.

And what waited at the circle's end? What was the culmination? "Sexual satisfaction +/- orgasms" was on the diagram, but in some versions it wasn't even part of the chart's main track; the physical, the carnal, didn't matter all that much. At the end was "non-sexual rewards . . . intimacy."

For Basson, such was the natural state of women's sexuality. She didn't make this case based on formal research; she'd developed her chart, she said, from her own clinical experience, and grateful patients had begged her to publish it. Yet while it seemed that her diagram might well represent the wan realities of many women's bedrooms, her assertion that she had drawn a picture of the inborn ignored the immediate genital reactions of Chivers's women, the overwhelming randiness of Wallen's monkeys and Pfaus's rats. She put forward a quaint and demure portrait, and strangely, stunningly, it was being adopted by the psychiatric profession—from the editors of the *DSM* to hordes of sex therapists—as though it were something wise and new.

One reason for this backward reeling was aesthetic and political. Basson's circle was supplanting a line, a diagram—credited to Masters and Johnson, along with psychotherapist and sexologist Helen Singer Kaplan—that had long been applied to both men and women, its progression going something like this: desire (first rather than laggard and nearly last) followed by physical arousal followed by pleasure. The line, the linear, could seem, from a certain feminist angle, phallic and patriarchal, decidedly unfemale in its symbolism, and at last Basson had provided an alternative, no matter that her lust-free woman was almost a Victorian paragon.

Another reason was bound up with a David-and-Goliath battle that some therapists saw themselves fighting heroically against the drug industry—against its rush to find, win FDA approval for, and market what was loosely known as a female

Viagra. Since the late nineties, when pharmaceutical companies had begun making billions by assisting erections with a chemical that affected the capillaries of the penis, the corporations had been seeking an equivalent for women. But this hadn't been going smoothly, because women's sexual problems usually aren't genital; they're entrenched in psychological complexities. Meanwhile, a set of clinicians had taken up a campaign, waged mostly within the psychiatric profession but also through the media, to make sure that the industry didn't manage to persuade huge numbers of women that they should feel more drive, that they needed a drug, soon to be discovered, to help them. The circle served as a useful emblem for the campaign, which was led by a New York University psychiatry professor, Leonore Tiefer, the author of a collection of polemics. Its title, *Sex Is Not a Natural Act*, amplified Basson's words, "We're just not talking about innate hunger." As for Basson's own attitude about the industry's search, she told me, "There are already enough date rape drugs around." Men would be sneaking lust pills instead of sleeping tablets into women's drinks to ease their assaults. Female modesty needed protection.

But maybe most of all, the circle was being consecrated as psychiatric doctrine because it gave sex therapists and couples counselors a solution to one of their most prevalent and stubborn problems—women's faint or non-existent desire for their husbands or long-term partners. The solution was low expectations. Clinicians had latched on to the diagram. They'd distilled it into a three-word lesson that they taught in treatment: "Desire follows arousal." They taught that arousal might take some time. Patience was a necessity; slowness and faintness were entirely fine; "lust" should be banished from the vocabulary. By lowering the bar, the circle offered therapists a standard for treatment that they might have a chance to meet.

And all the while, monogamy seemed to hover like an invisible angel above Basson's diagram. Occasionally Basson acknowledged that the new might be a key to combustion. But commitment, faithfulness, trust, familiarity—for her, these were the allies of female eros. Tenderness and intimacy ushered women along the circle toward the grand prize of yet more tenderness and intimacy.

Basson's colleague at the University of British Columbia, Lori Brotto, served on the *DSM*'s sexuality committee. It was a group of thirteen, and, with the manual about to be fully revised for the first time since the early nineties, she was in charge of the work on female desire. She had high cheekbones, a face that was all angles, hair cut fashionably close to her jaw. About women with the condition the current *DSM* called "hypoactive sexual desire disorder" or HSDD, she told me, "Sometimes I wonder whether it isn't so much about libido as it is about boredom." For her, monogamy was less hovering angel than grim reaper.

A psychologist whose sexuality research ran from hormones to acupuncture, Brotto treated women for HSDD in solo therapy and group sessions. "And unless you're talking about lifelong HSDD, which is rare, the impact of relationship duration is something that comes up constantly." In middle-aged women, she said, directing me to an Australian study that tracked hundreds of subjects from their forties onward, through menopause, hormones probably weren't as much of a problem as the length of time a woman had been with the same partner. (The Australian psychiatrist Lorraine Dennerstein, who had conducted that research, was more emphatic: "The sexual feelings of a new relationship can easily override hormonal factors.")

Yet Brotto, who was in her mid-thirties, who had been married for eight years, who was pregnant with a third child when we first met at a psychiatric conference, didn't mean to cast an all-encompassing pall on the ideal of long and loyal relationships. She was speaking about one aspect, about sex. And since monogamy simply was the prevailing standard—not only within the culture but within her profession—for success as a couple, and since it had a scarcely questioned status within the thinking of her committee, she was writing Basson's ideas into the *DSM*. They were ideas Brotto used with her patients, most of them long attached to one lover. She taught the circle, taught "desire follows arousal," taught these concepts as a way to begin to address disinterest in sex.

Seven years? Two? Less? More? Long attachment was impossible to define, turning points impossible to predict. But if Brotto could help her patients to become more responsive to the touches of their partners, if she could help them to feel more physically aroused, then even if they started out, in any given encounter, indifferent to their partners' overtures, they might reach a state of desire. To this end, she employed a little tub of raisins, passed around at her group meetings: six women sitting at a pair of pushed-together beige tables in a small windowless conference room. She asked each patient to take exactly one. "Notice the topography of your raisin," she instructed in steady cadences, her Canadian accent abbreviating some of her vowels. "The valleys and peaks, the highlights and dark crevasses."

Her career, her path to the raisin exercise and to her rarified spot on the *DSM* committee, had been mapped out by chance. As a first-year undergraduate, she knew only that she wanted to do research, no matter what the discipline. She hadn't thought about studying sex at all. "I grew up in a strict, Italian Catholic, don't-talk-about-sex environment." Even now, a silver cross hung from the rearview mirror of her car. She had knocked haphazardly on professors' office doors, hoping for anyone who would have her as an assistant. No one would; she was too young. But at last a professor invited her to help with his study of antidepressants and their effect on male rat libidos, so, for the next few years, she clutched a stopwatch and tallied copulations. Then, as she headed toward a doctorate, she steered away from animal research and toward clinical work, "because," she said, "the rat room smelled."

During her advanced training she did a stint with borderline personality patients. The condition mangles self-image to the point of horror: self-perception grows hideous. People are driven to cut or burn themselves; they ache to replace infinite despair with finite pain. Brotto's supervisor had developed a treatment that borrowed from the Buddhist technique of mindfulness. The idea was that keen awareness of immediate and infinitesimal experience, down to the level of breath or the heart's beating, might help to hold patients within the present and reduce their feelings of limitless torment.

While she was working with this supervisor, Brotto was also trying to help gynecological cancer patients with their sexual problems after surgery. The women who talked about lost libido, she thought, described their disconnection and sadness during sex in a way that was similar to the language borderline personality patients used to depict their entire lives. She wondered if mindfulness could help draw these women away from detachment and connect them to sensation.

She did some experimenting on herself. She didn't see herself as lacking in desire, but she did like to view herself sometimes as "an *n* of one," a single test subject for her ideas. Along with mindfulness, the treatment her supervisor had devised for borderline personality used cognitive therapy, with its stress on transforming patterns of thought, on reversing habits of damning self-assessment. And one day in yoga class, Brotto tried the combination.

As she arranged her body in her usual yoga poses, she attempted "a cognitive reframe. I said to myself over and over, like a mantra, that I was a highly sexual woman, a highly responsive woman. Not that I wasn't a sexual person, but now I was very consciously telling myself these things, taking on this persona. And there was the mindfulness. That's a part of yoga anyway; you're deeply aware of what your body is doing. You're aware of your breathing, your heartbeat. But that day there was a deliberate intent not only to listen to my body even more than I would normally in yoga but also to interpret the signs from my body as signs of my sexual identity. So my breathing was not just breathing through the pose; it was breathing *because* I was highly sexual."

Sensation and self-image became linked. She was in a tricky position, bent over and balanced on one foot and one inverted hand, when she had a profound moment. It wasn't that anything she was trying mentally was in itself so stunningly new. The power of positive thought was a cliché. And the acute concentration on the sensory harkened back to a style of sex therapy practiced by Masters and Johnson decades earlier. Yet by melding the two, something revelatory happened. Suddenly her straining muscles and racing heart were affirmations "of my sexual vigor, my arousability." She finished class and walked out onto the street and bicycled home with an exhilarating sense of her own body, her potency.

Brotto took what she learned treating borderline personality—the raisins came from that training—and what she discovered in yoga class, and tested it first with her gynecological cancer patients, then with a range of women who rued their weak desire. These days she sent her groups home to repeat over and over and over, "My body is alive and sexual," no matter if they believed it. And she guided them in the conference room, "Lift the raisin to your lips. . . . Notice that your mouth has begun to salivate. . . . Place the raisin in your mouth, without chewing it. Close your eyes and just notice how it feels. . . . Notice where the tongue is, notice saliva building up in your mouth. . . . Feel your teeth biting through the surface. Notice the trajectory of the flavor as it bursts forth, the flood of saliva, how the flavor changes from your body's chemistry. Notice the clenching of your jaw when chewing, the sensation of the raisin passing down the throat as it is swallowed. Notice the aftertaste and even the echo of the aftertaste."

Her results, published in the leading journals of sex research, showed her patients reporting stronger libidos and stronger relationships, though she was quick to note the caveats: that desire isn't easy to measure; that people are prone to claim improvement on questionnaires given by those who treat them; that just about any method that gets someone to think of sex can increase her interest in having it. And Brotto wasn't maintaining that she could grant her patients what they actually wished for. She had quoted to me from their files: "I want to have sex where I feel like I'm craving it." She sighed. She couldn't provide that, not without a semimiracle or someone new in the patient's bed.

I asked her about an irony within her *DSM* work: that while disorders were supposed to be abnormal, HSDD seemed to be a normal abnormality, a condition that was largely not psychiatric but created by our most common domestic arrangement. This was confirmed by all the women she met with who, she said, hadn't stopped desiring but who had merely stopped wanting, or had trouble wanting, their partners. Yes, she agreed, there was this tangle in psychiatry's reasoning.

She dwelled for a minute on the way our dreams and promises of forever seemed inevitably at odds with our sexual beings. "There is an element of sadness," she said, "when I think about the women I see, when I think about the couples I know, when I think about myself personally." She let out another sigh—or something akin to a sigh, a wordless note of grieving in a lower octave.

Leaning against the rail of his viewing tower, staring down at his monkeys and remembering the small cages that distorted the sexual interaction between females and males, Wallen thought that monogamy was, for women, a cultural cage—one of many cultural cages—distorting libido. He spoke about the research Brotto had mentioned: hundreds of women followed for fifteen years or longer, their relationships, biochemistry, desire relentlessly recorded. "The idea that monogamy serves the natural sexuality of women may not be accurate," he said.

Meana was sure that it wasn't. "I have male friends who tell me about new relationships. They say they've never been with a woman who's so sexual. They're thrilled. And I'm thinking, Just wait." Not only did monogamy not enhance female sexuality, but it was likely worse for women than men. There wasn't enough research on the topic, she said, but she talked about a German survey of committed relationships, showing that women felt desire wane more swiftly.

One reason for this, in her mind, stemmed from narcissistic need. Within the bounds of fidelity, the heat of being desired grew more and more remote, not just because the woman's partner lost a level of interest, but, more centrally, because the woman felt that her partner was trapped, that a choice—the lust-impelled selection of *her*—was no longer being made.

Like Brotto, Meana wasn't arguing against loyalty, against marriage. She alluded often to her husband; with adoration, she described his career as a professor of literature, a life she'd once wanted for herself. But when she discussed the work she did with couples, she made clear that she expected only very rare success in the realm of eros, if the measure of success was reviving lust. In around one-third or so of her cases, she could bring back something more mild.

Her method sometimes came down to scheduling sex, whether or not it was wished for, if sex hadn't been happening. She became a monitor, an enforcer. It was as though she were trying, almost brutally, to spade free something buried. "Fuck night," one of her female patients named it caustically. One of the married women I interviewed saw this kind of scheduling in a happier way. It was like exercising, she said, if you were one of the majority of people who would rather be reading or watching TV. By the time you left the gym, "with the endorphins going," you were glad to have been there, though you might not be anxious to turn around the next day and go back.

Therapists who claimed to restore lust on a regular basis, to instill desire in a high percentage of their patients, Meana thought, weren't judging their outcomes in any rigorous way, were deluding themselves, deceiving everyone. "This is big business—the books, the workshops. You could write a book full of promises every year, and every year you could have a best-seller."

She recalled giving, at a conference, a candid speech about her track record. One therapist, she said, approached her afterward with a common story. In sessions, a wife had suggested that if only her husband would be sensitive enough to help out around the house, she would want him in bed. So the therapist set him to work. She had him scouring pots. She had him tidying. She had him taking the kids to school and picking them up. But the sex didn't follow. "We tell men to water this little bonsai of women's desire," Meana said to me, "we tell them the bonsai has to be treated just so—and guess what?"

She wasn't objecting to men doing their half of the chores,

any more than her eye-rolls about intimacy meant that she didn't nurture, in counseling, empathy between couples. It was just that these things weren't likely to undwarf the tree's constricted limbs.

While Meana explained the problem with monogamy through her theory of narcissism, Sarah Blaffer Hrdy, a primatologist and anthropology professor at the University of California at Davis, raised evolutionary reasons. Her ideas challenged evolutionary psychologists who insisted that women are the less libidinous sex, the sex more suited to be monogamous. Hrdy had begun her career studying langurs in India. Within these monkeys, their jet-black faces surrounded by cloud-colored fur, the males commit rampant infanticide. They swoop in to kill newborns not their own. The same goes for the males in a number of other primate species. And female promiscuity among these types of monkeys and baboons evolved, Hrdy believed, partially as a shield: it masked paternity. If a male couldn't be sure which babies were his, he would be less prone to murder them. This insight didn't apply to all of our close animal ancestors; among the rhesus, the males tend toward caution and infanticide is seldom seen. Piecing together evolution's logic was an incremental process, full of incomplete patterns, causes that weren't universal. But Hrdy, with her theory of promiscuity as protective, added a compelling element within our ancestry.

And alongside this theory, she put forward an idea that might be relevant to countless species. It revolved around orgasm. Female climax—in humans and, if it exists, in animals—has been viewed by many evolutionary psychologists as a biologically meaningless by-product, a hapless cousin to male orgasm, with no effect on reproduction. Male nipples fall into this category; men don't give milk and don't need to for the sake of perpetuating humankind. The miniature look of the clitoris, compared to the penis, has helped to make the argument that female climax has no Darwinian importance, because the clitoris can have the appearance of a puny afterthought.

Somehow, this perspective has survived the recent mapping of the organ's underlying bulbs and wings. And the length of stimulation some women need to reach orgasm has reinforced the by-product argument; if the event had evolutionary significance, it wouldn't be elusive, unguaranteed. Especially during intercourse, it would happen more readily.

But the clitoral expanse—touched through the vagina—rivals the penis in total nerve-suffused territory. And as for the slowness of ecstasy, Hrdy flipped predominant thinking upside down. Her vision was a vivid example of substituting a female lens for a male one. Female orgasm could well be thoroughly relevant among our ancestors. Its delay, its need of protracted sensation, wasn't a contradiction but a confirmation of this; it was evolution's method of making sure that females are libertines, that they move efficiently from one round of sex to the next and frequently from one partner to the next, that they transfer the turn-on of one encounter to the stimulation of the next, building toward climax.

And the possibility of multiple orgasms compounded libertine motives. Another opioid rush—or a series of opioid infusions—might be in store with the next mounting. The advantages female animals get from their pleasure-driven behavior, Hrdy asserted, range from the safeguarding against infanticide in some primate species to, in all, gathering more varied sperm and so gaining better odds of genetic compatibility, of becoming pregnant, of bearing and raising healthy offspring.

Hrdy's stance on female orgasm as something probably much more than an evolutionary footnote was supported in another way. The data Pfaus had spoken about, of orgasm-like contractions in rats leading to greater chances of conception, matched fledgling, controversial evidence in women of climactic spasms guiding sperm up into the uterus. But even if female animals weren't actually having orgasms, weren't having the all-consuming subjective experience that we do, Hrdy's basic position about female pleasure held. Abundant stimulation was its own reward, reproductive benefit was its ultimate payoff, and in our near ancestors, this augered against monogamy.

Hrdy noted, too, proclivities in species more distant, polyandry from prairie dogs to sparrows. Or take the female of an arachnid called the book scorpion. Let her have sex with one male and, afterward, offer her that same partner. Forty-eight hours will have to go by before she's interested in mating again, though he is full of sperm and fully motivated. She seems wired to accrue an assortment of lovers and an array of sperm. Present her with a new male, and she is primed for sex within an hour and a half.

Meana, Wallen, Chivers, Pfaus, Brotto, Hrdy—all, in their different ways, from their different work in labs and observatories, in sessions of therapy and in the animal wild, pried apart assumptions about women, sex, constancy. Then there was Lisa Diamond, who began our series of conversations by emphasizing emotional bonding as the basis of women's desire.

Diamond, a professor of psychology and gender studies at the University of Utah, was a petite woman whose winning, raspy voice was always accompanied by big gestures. She talked with her hands, her shoulders, her neck, her dark eyebrows. When she and I first met, before a lecture Chivers had invited her to give to her department, she had just made herself semi-famous with a book titled *Sexual Fluidity*. It carried the academic blessing of being published by Harvard University Press. "In 1997," she wrote in her introductory lines, "the actress Anne Heche began a widely publicized romantic relationship with the openly lesbian comedian Ellen DeGeneres after having had no prior same-sex attractions or relationships. The relationship with DeGeneres ended after two years, and Heche went on to marry a man. The actress Cynthia Nixon of the HBO series *Sex and the City* developed a serious relationship with a woman in 2004 after ending a fifteen-year relationship with a man. Julie Cypher left a heterosexual marriage for the musician Melissa Etheridge in 1988. After twelve years together, the pair separated and Cypher—like Heche—has returned to heterosexual relationships." The opening went on to catalogue the sexual shifts, in both directions, of several more female figures, then asked, "What's going on?"

Diamond was a tireless researcher; the study at the center of her book had been running for more than a decade. Through long interviews and questionnaires, she'd been monitoring the erotic attractions of a hundred women who, at the outset, declared themselves lesbians or bisexuals or declined any label. From her analysis of the many leaps they made between sexual identities and from their detailed descriptions of their sexual lives, Diamond concluded that the direction of female desire was, above all, fluid. And after the book was published, she began collecting data among heterosexual women that helped to solidify her argument, that left her evidence less blurred by subjects whose sexuality seemed inevitably more likely to bend and transform.

Diamond, whose longtime partner was a woman, didn't claim that women were without innate orientations. But, she contended, female desire was generated—even more than traditionally assumed—by emotional entwining. Attachment was so sexually powerful that orientation could be easily overridden. Despite Diamond's provocative book title, in a way her thesis couldn't have been more conventional: closeness was almost all.

Yet something lurked, unaddressed, within her data on fluidity; her subjects weren't staying close to the same person. Relationships were being traded in periodically, and in the realm of sexual fantasy, they were being betrayed all the time. And suddenly, two years after our first meeting, when I mentioned the predicament of a woman whose story I will recount here in a moment, Diamond said, "In the lesbian community, the monogamy problem is being aired more and more. For years, gay men have been making open arrangements for sex outside the couple. Now, increasingly, gay women are doing it. It's interesting that lesbians like to call it polyamory, as though to stress love or friendship, instead of just letting it be motivated primarily by sex." She sounded almost like Meana; there was impatience with the veneer. As she continued, she turned to lesbian tastes in the X-rated, to "the difference between what's feminist-approved and what gets you off," to the doubtful presumptions that women need more narrative and more emotional meaning in their pornography, while men are more visual, more objectifying. "The stereotypes of male versus female, that male desire is far more promiscuous, seem more and more open to question."

Massage oil, a blindfold: the items Isabel had bought—hoping to alter the feeling of Eric's touch—when she ventured into the sex-toy boutique. On their visits, Calla and Jill weren't so reserved. Several months before, they had purchased a double-headed dildo—long shaft, two heads. Bodies arranged in the right way, they could penetrate each other.

These are four unions, four stories of loyalty and its limitations:

1

"Jill is more black-and-white than I am," Calla said about her girlfriend. "By personality, she's a jock. She's feisty. Things for her are either/or. I think maybe commitment comes more naturally for her. Once, maybe in our second year, when we were walking down the street, down the stairs actually, on Queen Anne Hill—there was this thick bed of ivy there—I started crying. I told her I'd never felt such unconditional love."

This was how she saw the woman she'd met four years ago in a lesbian bar, the woman she'd been living with now for a year. And this, that phrase, "such unconditional love," would reverberate later when I listened again to Meana, when she told me about an approach she took with only a few of her couples.

The bar had two levels. When their eyes had caught at a

distance—Jill standing upstairs and Calla below—Jill's had refused to let go. "Ballsy," Calla remembered, and recalled other impressions: Jill's sharp features, her combination of dark blond ringlets and green irises, the spareness of her athletic body, and the way that, when Calla had wandered off from their first conversation to flirt with someone else, Jill reappeared and announced, with whimsical flair, that she intended to compete. Calla took her home. For most of the year leading up to their meeting, Calla, who was in her early forties, had kept herself celibate in an effort to purge all the forces that had led to her last relationship, her last quick, eager pledge of fidelity, her last attempt at living together, her last disappointment, her last flight, her last repetition of this process, and that night with Jill, short, sinewy, brazen Jill, the sex went on ceaselessly, as though somehow a year might be pressed into hours.

For Calla, there had been a moment. One afternoon back in high school, in PE class, on a volleyball court two courts away from her own, with blue and white balls and black and white nets and cut-offs and gym shorts between them, she had noticed a classmate, a girl she'd seen and briefly spoken to before. But she'd never noticed her in this way, never had this reaction, this sense of invading chaos. Filled with dread, within days she gave herself a test. "I proceeded to go through in my mind the act of going down on her," she said. "And when I was done, I thought, No, I don't want to do that." To her great relief, this meant that she wasn't a lesbian.

Soon she was writing the girl poetry. Soon they were applying each other's makeup, telling each other how pretty they looked. She spent nearly all her nights at the girl's house, in her bed, the two of them in their underwear, tickling or running fingers along lengths of limbs. Things went no further. It wasn't until her freshman year in college that Calla stole away from a party, went to a dance at the university's LGBT center, wound up thoroughly immersed in a woman's body for the first time, and, in the wake of that night in that graduate student's bed, "realized how crazy girls made me."

Two decades had gone by since then. Cautiously she had put off living with Jill until the initial thrall subsided; meticulously she had tallied the pros and cons of how they were together; insistently she had promised herself that she wouldn't repeat the betrayals and disappearances of the past. The small apartment they shared was on Queen Anne, where she had wept gratefully on the ivy-ensconced stairs. Nowadays, after an evening out together, they might stand at the plate glass window that overlooked Puget Sound and share a rare cigarette and gaze out at the dark water, at the island's faint outline.

Sex sometimes began here, usually after six or seven or eight chaste nights. "Should we?" Jill would ask, inflecting her question with humor, with a sly reference to the count of nights that had gone by.

Calla would answer that they should.

"You don't sound too excited."

"Get into bed. Get out the toy and get naked."

"I've been forcing myself to push through my own resistance," she told me. "When Jill asks, it's like, I don't really want to, I should want to, I feel guilty about not wanting to. I tell myself I need to let go, that it's been too long. And then when we do start, it's playful, and I can feel her getting turned on, and that makes my body more focused. And meanwhile I'm fantasizing—it might be about other women, sometimes it might be about a man. Is there something wrong with me that I have to fantasize to be with her? I think maybe there is. I didn't have to at the beginning. Anyway, I orgasm pretty easily and so does Jill, and most of the time we orgasm again, and it is a release. And afterward my head is emptied out, and even with everything my mind was imagining I feel closer to her. So sometimes I ask her, 'Why don't we do that every night?' I say, 'We should do that every night.'

"Then a night goes by. Then another. I let them go, I make sure they go. I don't know why. And then the nights after that."

2

Susan wanted a low headboard. The master bedroom had banks of windows; she wanted the headboard to look right, not to block the panes. "And I wanted it to be good to hang on to during sex, which might have meant old-fashioned brass with bars, but that would have been too high. So I found a wooden one that went with a platform bed. It had these circle things, these circular openings, cut out of it."

The windows looked onto the suburban town where she lived with her husband. Below were their birch trees and the bird feeder he'd built for their son. At night, though, she recalled, "the windows kind of freaked me out. There were too many of them, and they turned into black holes of nothing. I think I must have been feeling something about my father. When he was dying, the hospice people moved him from his bed, which had a beautiful headboard by the way, with this blue silk upholstery, to a cot in front of a window that faced an air shaft." He was in his early fifties and single; he and her mother had divorced years before. "I was in college, and when I would come back to New York to visit him, I felt like someone was going to come in and snatch him there. I knew he was going to die anyway, but I felt like he was going to die sooner. He seemed so exposed next to that back window. I felt like it was stealing away his virility. It's funny, because there was another window in his apartment, a set of windows. And I remember nude sunbathers out there. They were on towels on a roof. That must have been east. The light that way was lovely."

With no transition, she said, "It was heartbreaking to lose my attraction for my husband. I couldn't talk about it. I didn't want to hurt him. And in a superstitious way, I felt like if I admitted out loud that it wasn't there anymore, it would never come back. I just prayed that it would. I get the feeling that for women it goes away more quickly than it does for men. I get the feeling that women are more dissatisfied than men are. It's the norm, but it's not talked about, and a lot of women struggle with the reality that they're not attracted to the spouses they're supposed to be with for the rest of their lives.

"We were very passionate in the beginning. But I think there's this whole misconception about women needing to be emotionally invested. I think it might almost be the opposite, that in the first part of a relationship the attachment is the product of the attraction. Sometimes, in long-term happy relationships, maybe, sex ends up serving the relationship, but at first it's the relationship that's serving the attraction.

"I don't know, though. Is that right? We were friends before anything else. It wasn't like I looked at him and thought, Oh, he's incredibly hot. It was the way he sounded. It was the way he smelled. It was the whole person. But I definitely found him really attractive.

"I remember one night our younger daughter came into our room. We'd just been starting to make love. I snuggled with her. I had no desire to be physically close with my husband. It had been like that for quite a while—that headboard never did get much use. She's a really good snuggler, and those windows were threatening. I could feel their presence. I'd had curtains made. In the winter they were heavy velvet. We did have sex maybe once a week, but it didn't reach me. My body would respond, but the pleasure was like the pleasure of returning library books.

"I had a friend who used to say, 'The longer you're married, the larger the bed you need.' And the thing about being repulsed by him was, I felt like my body was a room that I didn't want to mess up. Unlike that openness in the beginning when my body was a room and I didn't mind if he came in with his shoes on—when I wanted him to come in that way.

"He'd gained some weight, not a lot—I don't think I really noticed. And then I must have, on some level. It sounds crass. Maybe it was thirty pounds. You're taught that it shouldn't matter. He also started losing his hair. He's Jewish—black hair and dark skin and brown eyes. And I was very attracted to that. I'm freckles and fair. So he had all this nice black hair, and he started losing more and more of it, and it bothered me that he wouldn't do anything about it—he knew that I liked his hair, and he wouldn't use anything, and I felt like, I do all this stuff to try to look good, why can't you do that, too? He said that it shouldn't matter. And I said, 'Really? If I gained a hundred pounds you wouldn't mind?' And he said, 'I'd be worried about your health.'

"Somehow I lost my generosity toward him. I don't know how. It certainly wasn't just his looks. For women, it's not necessarily a beauty contest. Feeling generous isn't the same as feeling passion, but it can create a happier situation in your sexual life.

"I have a friend who told me about an article she read about how to heat up your marriage. One of the things on the list was having your husband jump you in the laundry room. She just laughed. 'My husband feels like my brother.'

"We never went to a psychologist until the end, when we were ready to divorce. I felt like seeing a therapist was only going to result in more tips like the ones I read in books—books written by therapists. We could try a hundred different emotional exercises. We could try new positions.

"So I just lay on that bed, holding my daughter. She truly is a gifted snuggler. It was like taking a muscle relaxant. I clung on to her and thought my morbid thoughts, She is the last physical intimacy I'm going to have before I die, she is the last physical intimacy I'm going to have before I die. And I felt those windows, even though I kept the velvet curtains closed."

3

Sophie and Paul's romance had begun when they were in nursing school. One night, ten years ago, a group of students had gone out to a bar and decided to play telephone. Paul sat directly to Sophie's right. "Sophie," she whispered to the woman on her left, "will you go out with me?" The question made it all the way around the circle, word for word.

They had been married now for eight years. They had three small children, the youngest under a year old; they both worked; and whatever time was left for them as a couple was swallowed by his studying and training for an advanced degree. Yet their bedroom seemed nothing less than anointed.

When she had first told her friends that she wished—wished badly—that Paul would ask her out, they looked puzzled. "Really?" they said. They thought of him as a dependable friend, not as the subject of dreams. But the man Sophie had just broken up with was a painter with a nipple ring glinting amid the muscles of his chest. He had done her portrait with dark flamboyance, depicting her as a corpse. It all seemed laughably melodramatic now, but for a long while she had been intoxicated not only by the Goth-style art, the gleam of jewelry, and the torso, but by the air of indifference—he rarely even bothered to brush his teeth—that seemed to keep women clustered around him always. He was unfaithful to her on a regular basis.

Then one day at nursing school, shortly after that relationship came to a cataclysmic end, Paul traded his light blue scrubs for a navy blue suit and delivered an assigned presentation. He was supposed to discuss an ethical dilemma that a nurse might confront, and he turned his task into a game of *Jeopardy!*, with himself as host and his fellow students earning points for posing the right questions to analyze and address the problem. He was animated; he came alive for her. She adored his ingenuity and eagerness; there was nothing nonchalant about him. Her memories of being rendered, in luxuriant brushstrokes, as a woman ready for burial started to fade. . . .

At the beginning of their first date—following the game of telephone—Paul steered over to the side of the road, stopped, bolted out, opened the trunk, and returned with a bouquet of roses, saying that he'd decided not to bring the flowers to her front door because she lived with her parents and the moment might be awkward. She was charmed by this hint of shyness and by the deliberateness of his buying the bouquet in the first place. They were deliberate together. Their dates lasted through entire nights, yet they postponed sex for two months and then made sure they weren't at her house or his—he lived with his parents, too. They planned the event. He booked a room at a nearby resort. When they finished making love for the first time—sex that was about as brief as she'd expected, given how long they'd waited—her eyes welled up.

He asked if she was hurt. She assured him she was not. He asked if she was disappointed. She told him that she was tearing up because she knew she would never have sex with a new lover again, and he, understanding that she was grateful, told her that the same was true for him. She did feel a level of regret that she kept to herself, an undertow of loss, but soon they were enmeshed again, and this time the lovemaking lasted, and over the next two years, until they married and moved in together, they conspired to find hours when they could have sex in their parents' houses without causing their parents discomfort, and the conscious intent involved in this collaboration, the plain acknowledgment of their desire for each other, the absence of all coyness about their feelings, taught them that a particular kind of magic could be created through simplicity and candor.

Boredom did not creep in behind this habit of transparency. Eros, for them, did not depend on suspense, on worrying if wanting was reciprocated. Some things weren't possible with three young children. The likelihood of the children's needs interrupting her and Paul's nights meant that she no longer slept naked, no longer had the radiating pleasure of feeling her nudity as a constant provocation. The eruption of the children's energy on Saturday mornings meant that those hours were no longer a time when her and Paul's desires could sprawl. And lately his training had cut their evenings apart. But distraction and fatigue didn't drain away lust. Their lack of guile somehow kept the attraction between them taut.

"We're really not subtle at all," she said. "My line is, 'Are you going to pay attention to me tonight?' Or he'll say, 'Am I going to get any action tonight?' And I'll tell him, 'Well, if you get off your study call and come upstairs before I go to sleep.' Or we'll agree to wake up at three in the morning."

She went on: "We never stop admiring each other. I'll say, 'You got your hair cut; it looks great.' And he still tells me all the time how good I look, even after the kids. 'Oooo'—this is one of his subtlest lines—'I love you in those jeans; can I get in them?' We make out in the kitchen. While we watch TV I'm touching him, or he's touching my breasts—even if there's almost no way it's going to lead to sex. I love that he loves to see me in these tight gray yoga pants I used to wear in nursing school."

Then, abruptly, she mentioned something hidden. She was a baseball fan, and when she had trouble reaching orgasm, or wanted to make love with Paul but felt that arousal was remote and needed beckoning, she tended to think about the Yankee's shortstop Derek Jeter. She smiled at the comedy of this confession. It was only sometimes that this extra help was required, she explained. "Jeter is the ultimate Yankee. Tall, all-American, everyone loves him—he's it. He comes home to me after winning the World Series. He's still in his uniform, and he throws me onto the bed and kisses me in a frenzy all over and thrusts right into me without me being really prepared for it. He just ravages me."

Yet even when she enlisted another man, she said, she felt little distance from her husband. It wasn't something they had ever talked about. "We've never asked each other. I don't think your partner needs to know. The fantasy is only a device. When you're with the same person for a long time, it's fine to use your mind to escape. I'm still with him, I'm still touching him. It's still *him*."

4

The woman in the zebra-striped cowboy hat lay on a blue blow-up raft at the shallow end of the swimming pool. Passie, watching her, was in her late fifties. The woman was on her back, one leg draped over either side of the raft and dangling in the water. Long, dark hair fell from beneath the black-and-white hat, a thin chain adorned one ankle, and, in between, her body was padded without being obese. "There were about twelve guys around her," Passie said. "She was nude. Big breasts. And vocal, because these men were playing with every existing part of her."

Four decades earlier, when the worshipfully maintained

mansions of her hometown were opened to the public, as was the tradition for one spring week each year, Passie had been selected to be a hostess. Once, great quantities of cotton had been shipped through this hub of the South. More than a century later, in the late nineteen fifties and early sixties, when Passie was in high school and college, the town poured its fragile pride into this annual display. She sat on one of the porticos. The crepe myrtle was in bloom: blossoms of purple and white and watermelon clouded the lawns and cascaded along the walkways. Her hoop skirt—pale pink—billowed. Her long gloves were dyed to match. "It feels surreal," she said now, "to have grown up in that time, in that place."

At twelve, in the Southern Baptist church where her father taught Sunday school and where she sang in the choir, she had stepped to the front of the sanctuary to have the minister press his hand to her head and lay her backward in the baptismal tank, to be saved. In her late teens, she pledged to meet the standards of the Little Southern Debutantes: to "at all times bring honor to herself" and to "be representative of the wholesome American girl." At the area's best women's college, she was taught how to swivel modestly in and out of an automobile, how to pause for a gentleman to guide her by the arm down a set of stairs, and how to pose in a group photograph if she was in the front row, with feet turned and hands folded and positioned to one side, so that the body suggested an elegant and demure S-shape below a straight, poised neck. "To this day I look at pictures and think that if women would just sit properly they would look so much better."

And during college, she was pinned. This was the pulse of

those four years. First, she would begin to date a boy from the state university nearby. Then he would give her a lavaliere with his fraternity letters to wear proudly around her neck. "Will you be pinned to me?" he would ask in the next step, and if she said yes, he would affix his fraternity's embossed shield to her blouse above her heart. About a week later he and all of his brothers would appear at the foot of her dormitory porch. She would come outside, and they would serenade her with their fraternity song: "And the moonlight beams on the girl of my dreams."

"I had a traditional vision of life, a fairy princess vision. My desire was for a Prince Charming who lived in a palace to sweep me off my feet. As a child, desire meant the wish for a new dress. As a teenager, it was wanting the right date for parties. At college, it was collecting the right fraternity pin and falling in love. You have your song and you dance at the parties on football weekends and you think he's going to be your husband. Lust didn't factor into it that much; it wasn't the driving force."

Nelson arrived as a blind date for her roommate, when, after her graduation, Passie was teaching French at a college a few states away. She had broken with convention in this way, choosing a career rather than marrying quickly, just as she had pushed against convention earlier: winning public speaking contests as an undergraduate and getting herself elected president of her state's Youth Congress, the first woman ever to hold the office. When the date with the roommate didn't work out, Nelson and Passie discovered, over Cokes, a little of what they had in common: loving theater (he sold silos for a living and acted with the town's amateur theater company) and classical music. "I found him an appealing person. And he was nice-looking. Not terribly handsome. But appealing. By that time I'd dated my share of men who were self-absorbed. He did things to make me feel special. I traveled with the foreign languages team, and if I was coming home late, he would have brought over food and left it for me in the refrigerator. He liked to leave my radio tuned to an Indianapolis station that drifted in from three hundred miles away."

As she recalled this, we were in their kitchen, Passie, Nelson, and I. Nelson sat in a leather wing chair, while she, on the other side of the counter, made a brisket for dinner and brownies for desert. Their home, near the college where she continued to teach and a few miles from the silo company from which he'd retired, was a single-story brick house on a leafy, trim cul-de-sac. The street might have run through a thousand American neighborhoods in a hundred towns and cities, the trees young, the smooth blacktop driveways outfitted with basketball hoops. Inside, Passie and Nelson's walls were decorated with landscapes: a nearby lake with a fisherman casting his line from a dinghy; a picket fence and horses bending to the grass of a field. Nelson wore a green golfing shirt tucked in tight over a loose middle and had a face and neck at once soft and strong, broad, generous. She wore a bright floral blouse and jeans that were slightly roomy over her slender, nimble frame.

About seven years ago, and thirty years after they were married, they were on vacation with their children and grandchildren, and while the rest of the family wandered through a fairgrounds one evening, the two of them went out to dinner at a favorite chain restaurant and had one of the few brutal fights of their decades together. She had already left their bed at home. At first, she had begun to sleep in the den sometimes because she had trouble with insomnia, but the separate sleeping arrangement had become permanent. Once, in the years before children, they had spent entire weekend days in bed. Later, after they'd started their family, if they were in the car together, just the two of them, she liked to read him the letters from his *Penthouse* magazines, turning herself on. But by the time they were in their fifties, she joined him in what had become his bed once per week, on Friday nights, joined him for what might be only a few minutes. He tried to thrill her, tried in all the ways they had, years before, learned together, tending to each other's bodies, listening to each other's skin. But the surface of her flesh seemed far off to her, let alone to him, and even a perfunctory orgasm had grown impossible. He came; they cuddled; she left.

And on vacation, her endurance disintegrated. All week she had felt entrapped by his wanting. With their children and grandchildren around them in their rented condo, she felt less than nothing in return. "This is not working," she erupted at the restaurant. "I know you're angry. *I'm* angry. If you come home one more week and say, 'Oooo, it's Friday night, you know what that means,' I'm going to leave the house. I'm not going to have sex with you anymore. I can't. I'm just not going to."

"I don't think I said too much," Nelson remembered. "I had felt this level of frustration in her for a good while, but we never talked about it. I knew something was going wrong, but I didn't know what to do."

Back home, they bought and read books of marital advice. Defeat followed determination. When Nelson heard an acquaintance say that he'd visited a clothing-optional hotel in the Caribbean, he mentioned the resort to Passie, half-jokingly, as a far-fetched idea that might rescue their marriage. "When he brought it up, I realized that I was interested—interested but very, very uncertain. I wasn't sure I could bare my body. I didn't know if I had the courage. No woman is ever convinced that she looks good enough to do that—definitely no woman of fifty-something. We thought it was just nudity, but they have lifestyle weeks when it isn't."

A month later, they were checking in for a weekend—in the front lobby, nudity was not allowed—and stepping tentatively from their room in bathing suits and, for her, a wrap.

"But even before I reached the pool, I threw caution to the wind. My bathing suit came off. I buried it in my tote bag. The guests were every age from twenty-five to eighty. There were women I tried never to stand next to, because they looked so good, and there were women who didn't look good at all. There were cesarean scars and hysterectomy scars and women who were totally out of shape, and I thought, If they can stand there and expose themselves, why shouldn't I? Bodies aren't perfect. The pool was up on a platform; you went up five or six steps to get to it. Every chaise lounge was filled with someone naked. There was a gal fondling someone's erection while she was having a conversation with someone else. There was a gal going down on another woman. And these men were rotating the float with the zebra-hat woman on the water, stroking her arms, kissing her breasts, stroking her legs, licking her clit. I spent thirty minutes watching her."

"I used to bemoan the fact that I'd never have sex with another woman in my life," Nelson said from his leather chair. As

Passie cooked dinner on the other side of the counter, he listed two or three of the sessions he'd had with women at the events they'd been attending, at hotels in surrounding states, every few months over the last seven years. His tone sounded vaguely rote, bewildered. It wasn't celebratory.

"I wanted to throw away my inhibitions. I decided she was going to be my role model," Passie said about the zebra-hat woman.

"Looking back, I think she had a stronger desire for other partners than I did," he said. "I think she felt it before that first trip."

"Subconsciously," she said.

On the table she set a basket of bread that their farming area claimed as its invention.

"We still have sex with each other." It seemed important, to him, that I know this.

"Nelson is my husband," she said. "I love him. He is the father of my children. When I say I love him, I mean it." She explained that at the events, she made sure he had someone to "play with" before she went off to another man's room.

"It's a paradox that I'm laying before them." Meana was speaking about a method she tried with just a few of her couples. Most of her patients weren't ready, she said; they didn't really want to take such risks. Her prescription didn't involve anything like alternative lifestyle gatherings. But it required a kind of divide. It meant the surrendering of safety.

She returned to a phrase, a dream, she had criticized

before: "You complete me." The seeking of a lover to embody these words; the pining for a love that will be unconditional; the search for a union that is absolute; the sense that our partners should give us what we were given—or what we believe we should have been given—by our parents; the craving for reassurance—*tell me I'm special, tell me I'm beautiful, tell me I'm smart, tell me I'm successful, tell me you love me, tell me it's forever, no matter what, till death do us part*—these were, for Meana, scarcely more than a child's cries. Yet most of us could not bear to give up on these longings. Most of us could not stand to relinquish the yearning for someone to be our fulfillment, our affirmation, because to turn away from such hope would be to acknowledge that we are, inescapably, navigating our lives alone, supported by love if we are lucky but, finally, on our own. Few of us want to navigate this way.

"There has to be an Other for there to be sexiness," she said. Yet in trying to save ourselves from our solitude, we strain to make our Others one with us. We flail; we grasp. We pray that selves will give way, that souls will combine. And eros, one of the forces we employ in our struggle, is crushed as we try to wring distance forever from our domestic lives. She wasn't suggesting that couples shouldn't turn to each other for comfort, for solace. "Love has to exist in different dimensions." Still, for most of us, in her eyes, something was out of balance: the longing to depend, to be propped up and protected, was given too much power.

With the couples who seemed willing, she liked to ask, "Why *should* she desire you?" or "Why *should* he desire you?" She demanded, "*Tell* me what's desirable about you. . . . And sometimes they look at me in a way that says, I can't believe you're asking me that. Sometimes they hear that as an insult, a slap. Sometimes my question hangs there for weeks. But slowly they realize what I'm doing. I want them to *focus* on what it is, to *know* what it is. I want them to work on what they see as desirable in themselves, to strengthen what they see as their strengths. And I want them to think about what they themselves wish for in a lover and try to turn themselves into that. I want them to make themselves *better*."

Her technique incorporated, too, tricks of disentanglement. Going out to dinner should begin with arriving at the restaurant separately. Date night should hold to the forms; it should mean a date. And chances should be seized to view the spouse apart. "If I can, I will have them watch their partner perform some function that has nothing to do with them. When I see my husband give a literary talk, and I'm in the back of the room, it's amazing how attractive that is to me. There he is in a way that has nothing to do with me, and my gaze gains a little bit of the gaze a stranger has on him."

None of this, she said, would lead to anything spectacular every time—or half the times—you made love as the years together accumulated. But sometimes, because the grasping had ceased, you might find yourself within a momentary, miraculous paradox, a brief merging after all. "It's about looking at each other in the midst of sex and feeling like you just dove into this pool of somebody *else*. It's about being astounded. It's about feeling breathless with that dive. That union. It's the fusion of two people with no differences in that instant. It's a complete I-am-yours-you-are-mine-I-don't-know-where-my-body-starts-and-yours-ends."

* * *

Had it only been Derek's failure as a player that deterred her from going to the Blazers' games, Alison knew she would have given herself a quick lecture and been there on the sidelines. Had it only been his chubbiness, his hovering as towel boy, and the irksome praise of the other mothers that had preyed on her, she would have told herself that the compliments were probably sincere, in their way, and reminded herself that her son was in fact a wonderfully spirited and open-hearted child. Though she might have felt that it would be nice to be the mother of the Blazers' leading scorer instead of the team's avid helpmeet, still she would have gone to the community center every Saturday—or many Saturdays, anyway—with hard-fought pride.

What plagued her, though, was that her minor issues with Derek and basketball mirrored her less ignorable issues with her husband and her life. She tried not to think about the parallels. And because she was a busy woman with a career as an editor that could easily crowd out other thoughts, she sometimes succeeded in keeping herself unaware. But she was also an analytic woman who reflexively drew connections. Thomas was pudgy and had never been much of an athlete, and Thomas's obsessive devotion to teaching the box-out to elementary schoolers was about on a par with Derek's alacrity with the towels—or would have been, except that Derek's role with the team was just one positive part of who he was, while Thomas's commitment to instilling basketball fundamentals seemed quintessentially Thomas. It seemed, more and more, to define him.

Until two years ago, while Derek was a player, her husband's faith in the character-building potential of his twelve basketball basics had struck her as maybe somewhat nuts but also admirable and poignant and slightly life-changing for the kids. But with Derek retired, the sight of her husband with his clipboard was a lot less sentimental. And his lengthy talk, over family dinners, about a new method for inculcating one of his twelve lessons—which might, for his players, carry over from the court and eventually help them to achieve thriving careers or happy marriages—made her feel that she might be serving a life sentence. Occasionally she imagined something devastating: that one day one of the other mothers would come over to her and pay her husband the same kind of cloying praise that she heard about her son.

As all of this was taking place at the community center and in her mind, two other things were happening. Thomas had purchased some stretchy bands, some dumbbells, and a video, and was putting himself through a persistent and hapless routine in the basement. And she and her husband were leaving their Manhattan offices once each week to meet together with a therapist, who liked to assign them exercises. In one, they had sat facing each other, palms resting on the other's palms, their breathing synchronized. Recently they had moved on to spooning, clothed, Alison behind Thomas, one hand on his heart and the other on his groin, or he behind her with hands positioned the same way, their inhaling and exhaling, the rising and falling of their chests, in gentle and exact alignment. They were supposed to let lust gather at its own pace, to postpone anything more sexual until desire coalesced within each of them, to feel no pressure to progress beyond this exercise, to understand that weeks might go by. They were supposed to simply experience the unity of breathing and allow this unity to permeate their hearts and genitals. But she, whose wanting felt extinguished and who was the target of the program, sensed no change aside from more and more futility.

This was the situation as she walked into the rec center gym clutching the hand of Derek's pretty younger sister. There, with about fifteen minutes before the jump ball of the season's first game, was Derek, giving a shoulder rub to the Blazers' captain. And there, farther along the sideline, was Thomas in a Blazers black jersey. This was something she'd never seen and hadn't prepared herself for, her husband wearing the top half of the team uniform above his jeans—most of the coaches wore polo shirts or sweatshirts, as Thomas himself always had—and it was jarring enough that she didn't see it precisely but perceived, instead, a semishapeless, almost blurry display of bloatedness and pallor: his shoulders and arms. She was already starting to remind herself that his misguided choice of shirt made no difference, that actually it was an endearing demonstration of caring about his and Derek's team, when she watched a mother step down the bleachers, taking long strides from level to level in her high-heeled, ankle-high boots, and stand next to Thomas.

His shoulders and arms, she recognized over the next seconds, and realized more fully over the following minutes, might be ghostly white, but they weren't puffy in the least. An outline of bulky strength was emerging. The woman in her suede boots, the mother of one of the Blazers' better players, started to chat with him, shoulder to shoulder, smiling. Whatever subject the woman raised—something about basketball, Alison assumed—it was plain that she spoke with affection. And as the minutes went by, it was clear that she was flirting with her son's basketball coach, with a sturdily put-together man who had been drilling sound principles into her child, lessons imbued with larger meaning.

Alison waited her turn. When it came, she pressed herself against her husband quickly from behind. She put one hand on his heart, told him where she wished the other one was and what she wanted tonight to be, and rejoined their daughter to watch the game.

CHAPTER EIGHT

Four Orgasms

Shanti, a former model who'd just turned fifty, took off her black boots, her black wrist bands, and her blue, red, and yellow Tantra Warrior choker. She slid off her dress, slid off everything, then arranged her body under a sheet and her blond head in the mouth of an fMRI cylinder. This was in Newark, in a Rutgers University lab with a wide glass pane dividing a pair of rooms. The giant cylinder was on one side of the window, and Barry Komisaruk, a Rutgers neuroscientist, and Nan Wise, a sex therapist and a doctoral candidate in his program, were on the other. They watched Shanti get settled.

Over the next hour, she would masturbate in various ways. She would use her finger on the external part of her clitoris. She would use a dildo to stimulate her G-spot and her cervix. Clitoral, G-spot, cervical—with Shanti and their other subjects, the scientists were trying to get clear and distinct pictures of the brain regions that burst into activity during three different types of climaxes. Komisaruk, a cheerful man in his late sixties with a horseshoe of curly gray hair, designed and made the translucent streamlined dildos himself to facilitate internal stimulation while avoiding contact with the clitoral exterior. He bought plastic rods, heated them at home in his oven, and bent them to his specifications.

Tantra Warrior was Shanti's self-created profession. She'd once been on the cover of *Elle*; now she made her living around Manhattan and the resort towns of Long Island, imparting erotic wisdom at soirees held by the erotically foiled, the erotically seeking. Komisaruk and Wise needed subjects like her who had no problem masturbating in public and amid the fMRI machine's bleating and clanging.

"When you're about to have an orgasm," Wise told Shanti through an intercom, "just raise your hand."

Shanti started on her clitoris under the sheet. Komisaruk, in khakis and a light blue button-down shirt, and Wise, in a crisp black skirt and silk blouse, were joined now by Wen-Ching Liu, a Chinese physicist and expert at interpreting neural imagery, in a white lab coat. They alternately glanced through the window and stared at a monitor on their own side of the glass, watching a map of Shanti's brain light up in constellated dots.

Komisaruk's decades of orgasmic research had begun with his wife's final stages of fatal breast cancer. They'd met at a summer colony when he was fifteen and she was two years younger. They'd gone steady right away and married five years later. At twenty-nine, she was diagnosed. She had just given birth to their second child. The metastases were swift and filled her with fluid and put her in such excruciating pain that she tore out her IV tubes and crawled across the hospital floor, trying somehow to escape her agony. "And I'm standing there like a dummy," he remembered, "unable to do anything."

His work at the time involved studying how sexual stimulation blocked pain in female rats, a tunnel of research that he'd branched onto after following a grand ambition, since college, to seek out the neurological underpinnings of consciousness. Watching his wife on her hands and knees, "I said to myself, I've got to do something useful." He would devote himself more thoroughly, he vowed, to understanding pain and figuring out whether sex might hold a natural analgesic. Could he distill an organic pain blocker to rescue sufferers like her? Along the way, after his wife died, his explorations with rats drew the attention of Beverly Whipple, nurse and sexologist and author of the early-eighties bestseller *The G-spot and Other Discoveries About Human Sexuality*. While he went on hunting for an analgesic on his own, he teamed with Whipple on experiments dealing with nerve tracks and women's varied climaxes, and that had led him here.

"Now we're getting it!" he exclaimed, eyes on the screen while Shanti worked. The clusters of dots were growing more dense.

"Wow!" Wise let out. "It's a Christmas tree!"

"She's moving fast," he noted, lifting his eyes fleetingly from screen to subject.

"For a Tantra girl," Wise said.

Shanti was imagining, she recounted later, "My lover touching me; him showing someone else how to touch me; lots of people watching; a line of guys waiting to stroke me, to lick me; then a cute, butchy girl putting her hand up my skirt."

"She's getting close," Komisaruk said. "That's the insula!"

Shanti raised her free hand.

"It's popcorn brain!" Wise said, inspired by the points of light.

But Shanti's session, it turned out, wasn't a great success. There had been some miscommunication, it seemed, when she'd been signed up as a subject. Erotic guru though she was, she told me afterward that she didn't think she'd ever had, in her life, a G-spot orgasm, and she knew she'd never had a cervical one. Her efforts with Komisaruk's homemade dildo didn't produce the data he was hoping for.

And then, too, he may have been overly optimistic in aiming to distinguish climaxes through brain imaging. In the months that followed, he didn't manage it, even once he had a set of subjects more versatile than Shanti. The needed machinery probably didn't exist yet, something he seemed both to have known and not let himself know as he leaped with scientific exuberance into the study. Brain regions could be glimpsed but not the terrain within and not the way those areas interacted. And the identifiable regions were broad, immeasurably complex. The insula—whose illumination had made Komisaruk's voice spring upward—was a neurological territory of pain as well as pleasure. When all his subjects had been through the experiment, Komisaruk could point to distinct spots in the brain that jolt into action with a touch of the clitoral exterior, the vaginal walls, or the cervix, but this was a long, long way from being able to separate out the almost infinitely intricate systems of ecstasy—systems encompassing much of the brain, from front to mid to back, from the prefrontal cortex to the hypothalamus to the cerebellum—in a trio of orgasms.

And that was assuming that the three different kinds of climax were a reality, that G-spot and cervical orgasms weren't a figment of popular suggestion and personal imagination. About the culmination of women's desire there was a swirl of uncertainty and a tangle of angry scientific and political debate, and it was all a reminder that in the twenty-first century it wasn't only the psychological questions of female eros that were unresolved but something seemingly much more basic: the mechanical workings of women's genitalia.

The array of plausible orgasms was a reminder, too, of Tiresias, who lived for seven years as a woman and informed Zeus and Hera that women are given the greater part of ecstasy.

The story behind Komisaruk's experiment traced back to Freud. The father of psychoanalysis, who made eros the essential substance of our psyches, decreed that stimulation of the external clitoris—he had no knowledge of the bulbs and wings—was like "pine shavings" compared to the vaginal "hard wood fire." A woman who relied on the clitoris for her orgasms was stymied, locked in an immature sexuality, thwarted physically and psychologically. Erotic womanhood was marked by orgasms through vaginal intercourse.

But Freud was hazy about one thing, a physiological problem that still bedevils the research of sexologists. He didn't deal with the dilemma that intercourse sometimes grazes, pulls, or puts pressure on the clitoris. Did he mean that mature, womanly climaxes were solely internal or was this external tugging and pressing acceptable?

It is impossible to know how many women attempted to train themselves to meet Freud's orgasmic standard, and which interpretation they took as the goal, but Marie Bonaparte—the same French psychoanalyst to whom Freud posed his question, "What does a woman want?"—was tormented by Freud's edict. Driven by her inability to climax through intercourse, and, it seems, interpreting the edict the second way, in the nineteen twenties she enlisted physicians to measure the distance between the tip of the clitoris—the glans—and the upper edge of the vaginal opening in their patients. She and the doctors collected, too, reports of the women's ecstasies. Then Bonaparte scrutinized the evidence. She concluded that her personal failure was due to the three centimeters that divided her key parts. Two and a half centimeters, she determined from her data, was the threshold; less than that and a woman stood a good chance of reaching bliss from a man's thrusting.

Next, Bonaparte consulted a Viennese surgeon. She had her clitoral ligaments snipped, her clitoral glans moved. Though the organ's nerves survived, the operation didn't achieve her orgasmic longings. Nor did a second try. She saw herself as doomed to what she termed "frigidity." But she kept on with her research, zeroing in on African women whose clitorises had been ritually cut, excised. Because of the loss of clitoral sensation, she asked, "Are African women more frequently, and better, vaginalized than their European sisters?" As a start toward interviewing subjects and finding out, she befriended Jomo Kenyatta, who was soon to lead Kenyans in rebellion against British rule, a war of liberation waged partly to preserve the Kenyan custom of clitoridectomy.

Bonaparte seems to have abandoned her African project without gathering much evidence either way, and by midcentury, scientific doctrine started to shift. Kinsey, from his interviews with thousands of women, and Masters and Johnson, from watching women having sex and masturbating in their lab, doubted the existence of the internal orgasm. Then, in 1970, feminist writer Susan Lydon published a clitoral manifesto. Men had forever "defined feminine sexuality in a way as favorable to themselves as possible. If a woman's pleasure was obtained through the vagina, then she was totally dependent on the man's erect penis . . . she would achieve her satisfaction only as a concomitant of man's seeking his." She proclaimed, "The definition of normal sexuality as vaginal, in other words, was a part of keeping women down, of making them sexually, as well as economically, socially, and politically subservient." But with the proper exaltation of the clitoris, "woman at long last will be able to take the first step toward her emancipation, to define and enjoy the forms of her own sexuality."

And soon the manifesto seeped into sexology. A kind of clitoral absolutism took hold. With her bestseller of the seventies, *The Hite Report on Female Sexuality*, researcher Shere Hite commanded an audience of tens of millions. She announced that the clitoris was the only locus of women's ecstasy. Whether from tongue or finger or the tuggings of intercourse, the external organ was where climax happened.

The absolute became accepted truth, imbued in popular consciousness. But in 1982, Beverly Whipple, Komisaruk's eventual collaborator, published her book on the G-spot. There was, she and her co-authors maintained, an area along the interior of the vagina's front wall that could bring on astonishing orgasms. She first hit on this phenomenon while working as a nurse with patients having bladder trouble. The zone could be elusive, she cautioned, and could be trickier to locate in some women than others. Sometimes G-spot climaxes produced ejaculations—not urine, she clarified, but a fluid that "resembles fat-free milk and has a sweet taste." She named the magical bit of anatomy after a German gynecologist, Ernst Grafenberg, whose forgotten writing from decades earlier, she discovered, had noted the same territory.

Grafenberg wasn't the first to have found it. A seventeenth-century Dutch scientist had documented the same region. But it was Whipple who brought it to prominence. Her book was translated into nineteen languages and set off an international firestorm. Critics railed that her research was anecdotal, flimsy, that she was sending women on an impossible hunt within the vaginal canal, a quixotic journey in search of superior, grail-like pleasure, that she was reviving oppressive Freudian ideals, that she was elevating patriarchal sex. The G-spot, her opposition insisted, was a fraud.

And nowadays, despite all the powers of contemporary science, the seemingly straightforward anatomical question, is there a G-spot? remains unanswered. The doubters view the phenomenon as a kind of psychosomatic bliss. They raise evidence like a study done recently by British researchers who sent out a questionnaire to thousands of pairs of female twins, identical and fraternal. If the G-spot exists, the scientists proposed, if it is a zone of actual flesh rather than an article of trumped-up faith, then identical twins, whose anatomies are nearly perfect copies of each other, will be far more likely than fraternal pairs to agree that they have one. The twin experiment had a classic structure, one that's been used repeatedly to separate the genetic from the learned, the objective from the subjective, in domains other than sex. And when the responses came back, the rate of positive answers was the same among the two groups. "What an Anti-Climax: G-spot Is a Myth," the *Sunday Times* of London declaimed. Women were now saved, one of the researchers said, from reaching for an orgasmic fiction and gaining only feelings of inadequacy.

But Whipple and Komisaruk, meanwhile, together and on their own, have accumulated data that leads to a different conclusion, with some of their evidence arriving through the orgasms of paraplegic women. In female rats and female humans, they've established that four nerve paths carry signals from the genitals to the brain. Two of these channels course straight up the spinal cord. But a third, the hypogastric tract, does an end-around; it doesn't join the spine till well above the pelvis, at about the level of a person's belly button. And a fourth, the vagus, whose name in Latin means "wandering," makes its wending way to the brain without relying on the spine at all.

Komisaruk and Whipple have shown the orgasmic importance of this multipronged map by working with women with severe spinal cord injuries, who, theoretically, shouldn't be able to feel what's going on below their waists. Their genitals should be insensate. And under examination in the lab, the paraplegics' clitoral glans have indeed proven dead. But the interior front wall of their vaginas and their cervixes have been plenty sensitive. As they masturbated by stimulating the wall or the cervix, the subjects reported having orgasms. The scientists validated their claims by gauging their sense of pain, taking their pulse, and measuring the dilation of their pupils. Sexology had already verified such readings as markers of climax: pain vanishes, pulse races, pupils widen. Whipple, sitting beside the masturbating women, collected the data, using a calibrated finger pricker and a pupillometer. And she and Komisaruk published papers arguing that the vagus and, in some cases, the hypogastric tracts were escorting the vagina's ecstatic messages around the point of spinal damage, while the signals of the clitoral exterior, by contrast, depended on the lower spine and were cut off. This, they reasoned, demonstrated that vaginal orgasms were real and distinct, that they weren't merely due to oblique pulling and pressing on the external clitoris. And, they explained, the two circuitous tracts, the hypogastric and the vagus, were why healthy women described vaginal climaxes as feeling different from the external, the clitoral, as feeling "deeper," more "throbbing," "stronger." Somehow the less linear, more sinuous and imbedded nerve paths created these sensations.

But even for those who dismiss evidence like the twin study and who trust the lessons of the paralyzed and the truth of the internal orgasm, a primary mystery clouds everything.

What is the exact anatomical origin, or blend of physiological sources, for this variety of bliss? Is the G-spot a spot or a diffuse and even slightly morphing province? Is it an entity of the vaginal wall itself or is it more about what lies behind the wall, the nerve-dense clitoral extensions, the wings, charted in the late nineties? If it's about those extensions, and the stimulation they might receive through the wall during intercourse, are vaginal orgasms clitoral after all? Or is Komisaruk right in deducing, from his data with paraplegics, that this probably can't be the case, because in these paralyzed women the nerve tracts from the extensions would be severed just like the paths from the glans?

And how should anyone grapple with understanding the mechanics and nerve routes of the proposed third type of rapture, the cervical climax, a late addition to the orgasm debate—and one with a possible reproductive relevance? As with rats, stimulation of a woman's cervix facilitates a hormonal release that can, to an unknown degree, aid the fertilized egg. But finding out, scientifically, whether women can actually have a cervical climax may be impossible. It's difficult to imagine how the experience could be isolated, difficult to imagine the dildo, kitchen-made or otherwise molded, that could completely bypass stimulation of the walls and touch only the back of the canal.

Setting themselves to unclouding the G-spot if not the cervical uncertainties, two French doctors lately positioned a woman, who said that she had vaginal orgasms, in gynecological stirrups. They had her lover slide himself inside her, and they put a sonogram's scanner on her pelvis. This vision of intercourse revealed that a pair of the clitoris' underlying projections might be the solution to the G-spot puzzle. These projections embrace the spongy, nerve-lush lining of the urethra. And on the sonogram, when the penis struck a particular zone on the front wall, the extensions were stirred into a scissoring motion, massaging the urethral lining. This, one new theory went, stoked the lining into an overload of neural activity—and the woman into climax. So the spot was the source of the scissoring, and the ultimate origin of the orgasm was the urethra's cushiony outer layer.

Komisaruk and Whipple have released a guide for the general reader: "If one or two fingers are inserted into the vagina, with the palm up, using a *come here* motion," the zone can be found. "Women have reported that they have difficulty locating and stimulating their G-spot by themselves (except with a dildo, a G-spot vibrator, or similar device), but they have no difficulty identifying the erotic sensation when the area is stimulated by a partner. To stimulate the G-spot during vaginal intercourse, the best positions are the woman on top or rear entry. The orgasm resulting from stimulation of the G-spot is felt deep inside the body."

None of the efforts on either side has put an end to the vaginal versus external disputes. Nothing seems likely to. About half of all women believe they have a G-spot; half think they don't. But Komisaruk and Whipple, using their finger pricker and pupillometer, have verified something that transcends anatomy, something that hasn't brought much doubt: there are women who can think themselves to orgasm with no touching whatsoever. For reasons unclear, it's a capacity much more common in women than in men. In Komisaruk's and Whipple's lab, imagining lovers or, for some, passages of music, women have sent themselves into ecstasy.

One afternoon I watched as Wise, Komisaruk's associate on the fMRI study, lay back in the cylinder and demonstrated. It was all about breathing, she told me before she went into the machine, and about the strength of the pelvis and about "knowing how to circulate the energy." She kept her preferred fantasies to herself.

I asked if it was truly an orgasm.

"There are all kinds of sneezes," she said, "but there's no question it's a sneeze."

Now she was motionless under the sheet. On the screen, the constellations of dots were getting thicker and thicker, more crazed. Five minutes and nineteen seconds after she began, she raised her hand.

CHAPTER NINE

Magic

Martina Miller, the coordinator, counted out tablets. Wendy filled out a questionnaire. She prided herself on efficiency, and she was efficient at this. She sat at Miller's desk, facing photos of Miller's carefree dogs in magnetized frames on a file cabinet. She removed the paper clip that fastened the questionnaire's many pages, swiftly read each query, quickly checked the boxes beside her answers, straightened the pages by tapping them sharply—*click-click*—against Miller's white enamel clipboard when she was finished, reattached the fastener, and passed the document back to the coordinator.

In return, Miller handed Wendy a new supply of medication. Red slacks, a canary yellow scarf with orange trim—Wendy radiated bright hues and optimism. She said thank you, gave a split-second's giggle, zipped the pill jar into her glossy shoulder bag. But there was a glitch. Checking her computer, Miller pointed out that Wendy had been missing some of her reports, that she hadn't been making an entry in her online diary every time she put a tablet on her tongue.

"I know, I know," Wendy confessed. "It's a mess. I keep forgetting." For two or three minutes, her upbeat armor cracked. There were no tears, only fear expressed in cheerful tones, as she stopped in here at a center for sexual medicine in a Maryland suburb. Soon she would be outside, in her car, in the sun, away. She would be driving through the May afternoon to her ten-year-old daughter's lacrosse practice. But now she explained to Miller that she'd taken the drug, felt nothing, done nothing with her husband, fallen asleep and ignored her diary the next day, with only failure to record. She hoped her first round of pills had been placebos.

Answering ads on the radio, in newspapers, on Craigslist, the women had arrived to enroll in the trials all fall and winter. I'd watched that stage of the process at another clinic, near downtown Washington, DC. The tiny drug company, Emotional Brain, had enlisted centers all over the country, clinics run by psychologists and gynecologists and everyday physicians, some taking part just because medical trials were a facet of their practice, others because they believed that EB's inventions, Lybrido and Lybridos, might prove distinct enough from the earlier chemicals of other companies, might be ingenious enough in their composition and precise enough in how they would be prescribed, to be the first aphrodisiacs to make it past the FDA, the first to give doctors something, something reliably successful and government approved, to offer women like Wendy.

"They use terms with real emphasis, words that are violent," Andrew Goldstein, who ran the DC center, said about his patients. The light in his office was soft. A close-up photograph of a cherry tree hung opposite his diplomas. "This is like someone cut off my arm; this is not how I see myself; this is like something's been ripped away from me. Stripped away. Stolen." He was among the most prominent gynecologists in the country, the president of the International Society for the Study of Women's Sexual Health. And he was all but exultant. He didn't stand to profit financially if the data from the trials panned out, if the two drugs outperformed the placebo, if the side effects were mild, if the FDA gave its blessing. He'd signed on for trials of other medications, molecules aimed by pharmaceutical giants at the same despair, the feeling of desire's vanishing, aimed at the same market, worth over four billion dollars a year in America alone. Then, for the past two years, he'd taken a hiatus, out of frustration. But Lybrido and Lybridos had rekindled his hope. He sensed solutions. And it wasn't only that. EB's diagnostic method, its gleanings of the genetic and the learned through blood work and interviews and its algorithm that compiled and processed these gleanings, would allow new glimpses into women's sexual brains.

"The tools we've had up till now have been like flint knives." His field's wherewithal, for comprehending, for treating, had been blunt, crude; it had belonged to the Stone Age. As we talked between his screenings of possible subjects, he wore a blue and white pin-striped shirt, a white lab coat. His voice was scratchy and high. He had a cherubic face and full gray hair, so that he looked sometimes childlike, sometimes stately. The gray seemed to disappear when he spoke about EB's algorithm, its pills. "God bless! This is fine-tuned!"

If EB turned out to be on target, he said, there would be fantastic changes, specific, vast. He would have a drug to help a subset of his patients, women whose antidepressants suffocated their desire. He would have a way to understand one of the conundrums of his field: why birth control pills snuffed out sexuality in some—but far from all—women. He would have something much more accurate than the current blurry grasp of testosterone's effect on female libido. And more than anything, for all sorts of women, he would be able to restore what they felt had been torn away.

An African-American law student who, after five years with her boyfriend, couldn't trick herself into the wanting she'd once felt, could only trick him. "I use a lubricant, so he doesn't know," she told Goldstein, as he interviewed her for EB's pools. A divorced mother of three who sensed herself, with her lover, slipping into a sexual indifference that was familiar from the demise of her marriage. "When we split up," she said about her ex-husband, "it was like going through a second puberty. So I attributed what had gone missing to who he was." She gave some attribution as well to her children, the energy they drained away, the physical and occupational therapy appointments her disabled son needed each week. But with the indifference returning, she was starting to doubt those attributions, starting to wonder if it was something about herself. A bank officer who, answering Goldstein's questions about her past, mentioned where she'd met her husband. "It was at Nashville International Airport."

"How'd you meet him at an airport?" This sort of detail didn't matter at all to EB in deciding who to enroll in its trials, but Goldstein was that kind of doctor. He liked to get to know the women who sat across from his desk, even if they weren't his patients, even if they would only be in and out of his office a few times over several months, to pick up tablets and answer follow-up questions, even if they would be gone forever after that.

"I was a screener," the bank officer remembered. "I was in college, and I was a part-time screener. I was coming back from lunch in my uniform, and he was looking at me, and I said, It's not polite to stare at a woman without saying hello. I turned around, and he followed me."

"Obviously he had something to say."

"Obviously," she said, and she and Goldstein laughed together.

"How long did you date?"

"It was extremely fast. June we met, March we married."

And for years, even with young kids, she'd felt that speed, that sense of something predestined; she'd counted on the rush of their combined bodies. Now, in her late thirties, all happened slowly, all waited at a receding distance. Often she faked her orgasms.

"When he initiates sex, do you feel anxious?"

"I do."

"Stress?"

"I try not to show it."

For every woman who wanted to enroll, there were a range of reasons. There were the demands of law school, the disabled son, a self-consciousness about added weight, a fibroid surgery that seemed to have caused damage, though a neurologist could find no loss of sensation. "When he plays with me, when he tries to jump start me down there, I don't feel it, I don't understand," the bank officer said. "That's why I need to be in this study." There were lots of factors, always, Goldstein told me. But as I listened, it sometimes seemed there was only one. There was no ripping away, no theft; nothing violent had occurred; there was only a leaving behind. Time had passed. Desire was back there. That was all. That was violent enough.

Lybrido, Lybridos, the pharmaceutical efforts that had come before them, the inestimable millions or billions that the industry had poured into research—the race was for a drug to cure monogamy. This was the main demand, the market with the biggest potential payoff.

"I just want to know," one woman asked at the end of her interview, after describing the man she'd spent the past seven loving years with, "is this medicine going to work? Am I going to get my freak back?"

In her front yard, one May evening two years ago, Wendy sat with her neighbors on lawn chairs. Behind the women, their houses stood quiet, modest, in matching brick; beside them was a portable fire pit on iron legs, flames breathing heat into air on a cusp between spring and summer. Their upstairs windows were opened wide, letting that air transform the rooms. Their children swooped on a rope swing behind Wendy's house; their husbands were at an Orioles game; the women sipped their wine.

A beeper went off, faint, then louder, more insistent, bleating into the suburban calm. Wendy's next-door neighbor sprang up. The study Wendy and two of the others were in that year worked a bit differently than the EB trials; electronic diaries, logs of sexual acts and feelings, were to be updated every day, and Wendy's friend sprinted inside to silence the company's automated reminder before it began screaming across the neighborhood. They were taking Flibanserin. They tracked their responses for the company and talked with each other—and with their other friends at the fire pit, who were monitoring their progress—about whether the experimental pill was working. They wondered together what the odds were that two or all three of them had been given the placebo. They agreed, on evenings like this, or in the mornings, over coffee after putting their kids on the school bus, that whatever they'd been handed wasn't having any effect, though one of them thought there might be a chance she was starting to feel something.

Intrinsa and Libigel, Flibanserin, Bremelanotide, these were among the defeated drugs that had come before Lybrido and Lybridos. Intrinsa and Libigel, a patch and an ointment, delivered infusions of testosterone—and within testosterone's failure with the FDA were lessons about how little science had managed to sort out when it came to the biochemistry of women's lust.

Somehow, by mechanisms still just broadly understood, testosterone primes the making and messengering of dopamine, the brain's courier of urgent wanting. This priming happens within and right near the almond-sized hypothalamus, which sits down by the brain stem and helps govern our base drives and bodily states—hunger, thirst, lust, body temperature. Intrinsa and Libigel tried to influence the dopamine circuits that are devoted to sex by sending more testosterone through the blood to the brain.

Spiking dopamine directly, instead of using testosterone, can cause trouble. The techniques aren't refined; the results can be a brain in overall overdrive, damage to the circuitry of motor control, severe nausea, a risk of addiction if you spike too often. And, Pfaus told me, testosterone might assist desire in ways that reach beyond dopamine by tweaking other crucial neurotransmitters. Given all this, a drug supplying extra testosterone seemed a promising approach. But there were baffling complications. They were known, to some extent, even before the testosterone aphrodisiacs went into development and into trials. Whether because testosterone isn't the main primer after all, as some scientists argue, or because there is too much other biochemistry at play, the puzzle was this: add testosterone to a woman's bloodstream, and you wouldn't necessarily cause a rise in desire; deplete the hormone, and you wouldn't dependably reduce libido.

Oral contraceptives, Goldstein said, launching into a lecture on hormonal confusion, could all but eradicate a woman's blood-borne testosterone. "Birth control pill–takers have free testosterone levels one-tenth, one-twentieth of where they would normally be." This situation hadn't always been so drastic. Pharmaceutical companies had lately been fabricating contraceptives that pushed testosterone lower and lower to strengthen a sales-enhancing side effect—the elimination of acne. For plenty of women, the hormonal decimation didn't seem to make any difference to desire. For some, the pill generated drive, probably, Goldstein went on, because women without worry of pregnancy, with lighter or less frequent bleeding, were more likely to seek out sex. But for others, oral contraceptives led to a crash in libido. Why were some women harmed by the bottoming out of testosterone, others unaffected?

Menopause added to the riddles surrounding the hormone. Middle-aged women and lots of their physicians tended to blame menopause for dissipating desire. Doctors gave out testosterone as a remedy—they gave it in a way known as "off-label," unapproved by the FDA, semilegal. And some women reported successful results. Yet despite popular belief about the time of life when the hormone dropped, menopause didn't actually bring a decrease in testosterone at all; instead, there was a slight rebound. In truth, a steady decline had taken place long before, when a woman was in her twenties and thirties. And the depth of the decline was no worse than what went unnoticed in uncountable women who took the pill.

Was there a way to make sense of any of this? Was there a way to draw tight links between the physiological—whether something as straightforward as a hormone count or as complex as menopause—and libido? With estrogen, possibly. Around menopause, loss of estrogen led, in some women, to dryness that could undermine desire—even though, if you hooked these women to a plethysmograph and played a pornographic movie, blood raced as it did in far younger subjects. The tissues just weren't manufacturing as much fluid anymore when the blood flowed in. So the psychological pathways of desire were intact, but the chemical reactions responsible for wetness were impaired. And the tissues themselves could thin. This could lead to obvious problems: if intercourse was uncomfortable, you weren't likely to want it; if it was downright painful, you would probably avoid it; either way, you might quit thinking about it; desire might be destroyed. Then again, something else was obvious, too: there were any number of other ways to have sex. But a deficit—immeasurable, maybe immense—was at work. Your mind wasn't going to be hearing the messages of your genitals as well as it once had. And the communication could be tenuous to begin with. Chivers's experiments had shown this; her subjects could seem deaf to what their genitals were saying. Lubrication was part of the language—with that diminished, the lustful messages might be more muted, the mind less prone to the awareness of desire, the brain and body much less easily caught up in a loop of yearning.

Yet there was the Australian study: on desire, new relationships trumped menopause, easily. And Goldstein spoke about how readily dryness and atrophy could be treated with estrogen supplements, with low and safe doses. Lubrication was restored, tissues regained health, but libido didn't revive reliably. Stubbornly, the effects of hormones evaded logic. Sometimes desire seemed to hide itself from science.

He returned to talking about testosterone. He was among the thousands of doctors who provided it off-label, flouting the spirit if not the letter of the law. He didn't hesitate to discuss this. He felt he had to do what he could for his patients. Women came to him after being dismissed by their family physicians, by other gynecologists. "If I had a nickel for every time a patient has said that her doctor told her, Just have a glass of wine." He gave the hormone to women of all ages, though not indiscriminately—he used his own criteria, his own intuition, to try to figure out who it might benefit. He looked for low blood readings of testosterone, poorly predictive though he knew these to be. He weighed the histories he heard during his interviews, listened for the disappearance of erotic dreams. This, in his mind, was a telling sign: the evanescing of sex from the life of the unconscious. He dwelled on the clues he gathered, followed his hunches. He guessed that by proceeding in this way, he helped more than half the women he provided with the hormone. But that left a lot of his patients inexplicably unreached and a lot who didn't qualify, by his calculus, for this treatment. And it left him reading dreams, practicing medicine by a system that was barely systematic.

This muddle, this imprecision, this inability to predict, lay within testosterone's latest struggles with the FDA. In trials with a thousand women, a pharmaceutical company had collected data to back its product, Libigel. The product did seem to work—somewhat. On average, it made desire rise—undramatically. And yet in the trials, a fake gel, a placebo, had aided libido as much as the medicine. Self-persuasion seemed as potent as the drug.

Around the fire pit or after the school bus, the chatter about Flibanserin was light, filigreed with a joke or two, the way Wendy liked talk to be. Yet with her friends gone, after wine or coffee, she felt something insidious, a helplessness, a foreboding, a sense that she would be unable to protect—to protect what? Not her marriage, not quite that. She trusted that she and her husband would remain together. It was, she said, love that needed saving. It was—she used the simplest of words—"happiness." She wrestled to prevent desire's further and further withdrawal.

After college, she'd met her husband at a sports bar, laughed with him over a foosball table, laughed more later that night as he clowned, concocting his own dances. This was in New York, where she lived for a few years, intending to return home to the Midwest, to marry there, to build a life close to her family. But she found herself unguarded with him, without need of hiding, which was new. She admired "the way he could make fun without making anyone feel bad." And there were moments almost too inconsequential to be described. They had gone to a video store one evening, stood in a crowd waiting for the clerk to set out a batch of movies—and when the film she wanted appeared, she paused before reaching, to let someone else take it. He said that he liked what she'd done, and so many years afterward she still recalled his plain and half-shy praise, the pleasure he'd taken in her gesture. Gradually, as they dated, she had grown entranced by him, thinking, when they were out with other couples, I just want to get home with him, I just want to get home with him. And then they had created a home together, inside the three-bedroom brick colonial, a life she had never regretted. It was only that she was scared.

Years ago in their house, she had seized his hand and hurried him up the stairs. Now she waited, somewhat like prey though the predator was tender, though he was cherished. "He'll move closer to me in bed, or put his arm around me, or rub my back." Once a week he tried to reach through the invisible barriers she built; once a week she tried not to refuse him. And like an indestructible machine, she climaxed regularly when they did make love, as she always had. But the next night she returned to being the person she'd become, the woman who willed herself to sleep or focused intently on her book as he climbed the stairs. It was impossible to understand, how those stairs had changed.

Unlike Libigel, Flibanserin tinkered directly with neurotransmitters, but its tinkering was too delicate. In trials, it didn't do enough good to get past the FDA. Wendy and her friends had been accurate barometers. Other medications had other troubles. A few years before Wendy's group got involved in such studies, a drug had arisen through happenstance. A team of University of Arizona researchers had been exploring a chemical as a sunless tanner, a compound that would fuel a set of pigmentation-producing cells in the skin called melanocytes. But when the scientists ran tests on a small number of men, they heard back from almost all with an unexpected response: sudden and stunningly rigid erections. And unlike the effects of Viagra, which were all about the hydraulics of the blood, the bronzer tilted the mind, left it reeling with lust. Viagra bestowed hardness where there was drive; the tanner, the researchers found, bestowed both.

No one was sure about everything the chemical did in the

t browned the body, but with each dose, the me-
area of the hypothalamus, part of Pfaus's "ground
e," sent extra dopamine coursing, for several hours, through gray matter. Not only did the appetite for sex shoot up, but the appetite for food was killed off. This fit with a known interconnection inside the subregions of the hypothalamus, with a relationship between the basic motivations: sex, food, sleep. If the desire for one gets overwhelming enough, the others stop mattering.

The company that bought the rights to the chemical believed it had something remarkable. It sculpted the compound, culling out what tanned and carving away what deadened the appeal of food, saving those effects for drugs to be developed later and concentrating first on sex. For a preliminary forecast of how well the medicine, christened Bremelanotide, might do, the company sent a box, via FedEx, to Pfaus, who sent the new molecule, via mini-skullcap, into his rats. The males sprouted erections at an extraordinary rate. This was good news for the company, given that Viagra and its chemical cousins don't work for around one-third of impotent men. But what inspired corporate glee was the female reaction. The tallies of hops and darts, of head-pointings and prancings away, of climbing the hindside of a male and doing a demonstration hump—the utterances of *I crave you* in feminine rat parlance—soared.

Next the company turned to hundreds of women who lamented the state of their libidos. "I was one hundred percent into it." They recorded their experiences in trials after absorbing Bremelanotide by way of a one-dose nasal inhaler. "I was tingling and throbbing." "I was focused on sex; I wasn't thinking about anything else." "My climax was like it used to be." "I was able to climax multiple times." At the Maryland sex clinic where Wendy now got her EB pills, the psychologist in charge had also had high hopes for Bremelanotide. His center had taken part in those studies, and he remembered one woman inhaling the chemical, then sitting in the waiting room until she could have her vital signs checked for negative side effects. Overcome by metamorphosis in mind and genitals, she declared to everyone in earshot, "I've got to call my husband to make sure he's home when I get there."

The signs for Bremelanotide were spectacular. A major magazine put the drug on its cover with an illustrator's vision of midtown Manhattan. Taxis had screeched to a halt. An orgy raged on hoods and windshields, on the roofs of buses, on the pavement of a traffic island.

But the snag, the Maryland psychologist recalled, was that some women weren't celebrating in his waiting room; instead, a few were in his bathroom, vomiting in the stalls. Besides the bouts of nausea, blood pressure jumped in a small percentage of subjects. About halfway through the FDA process, with tens of millions still to spend on more trials, the company slunk away from its application, knowing it would never get approval for an aphrodisiac with those hazards. It had since moved on to studying an intravenous version, which didn't seem to bring on queasiness or hypertension, though how many people would be willing to stab themselves with a needle for the sake of desire was a source of doubt.

And the company had always fretted about something else. In the initial phases with Bremelanotide, after seeing the randiness of the female rats, the euphoric reports pouring in from women, and the orgy on the magazine cover, company officials got frightened even as they were overjoyed. At meetings, Pfaus remembered, they anticipated that the drug might be too effective for the FDA, that the cover image of women splayed feverishly on cement, their legs hooked around strangers, would haunt the agency and scare it off. There was no telling whether the FDA would have raised the specter of sexual mayhem had the application reached a conclusive review, but the company huddled with researchers like Pfaus to ask if there were any data to suggest to the agency that the chemical's impact would be "selective," that Bremelanotide-sniffing wives and daughters wouldn't "want to go off and do the football team."

This resonated with what Goldstein recounted from his involvement with Flibanserin. In Flibanserin's trials, he hadn't taken his usual outsider's role, interviewing women, dispensing medication. He'd been hired as an advisor by the corporation that owned the molecule; he'd been in on strategy sessions. "When you're going to the FDA with this kind of drug, there's the sense that you want your effects to be good but not too good," he said. Too good hadn't turned out to be Flibanserin's problem, but, he explained, "There was a lot of discussion about it by the experts in the room, the need to show that you're not turning women into nymphomaniacs. There's a bias, a bias against—a fear of creating the sexually aggressive woman. There's this idea of societal breakdown."

With her yellow and orange scarf wrapped under her chin, Wendy told the coordinator—who was keeping information up to date after checking about the missing entries in Wendy's EB diary—that she seldom fantasized about other men. Even passing images were rare. "I'm very attracted to my husband," she said to me, a steely undertone just scarcely audible in her chipper voice. It was the kind of answer I'd heard from some, though far from all, of the women I'd spoken with, as if their feelings for their partners needed safeguarding, were better left unbetrayed, even in their minds. They seemed to adhere, consciously or reflexively, to timeless rules about the way women should and shouldn't be. Did this take its toll on the sexual circuits of neurotransmitters, which, like all our circuitry, can be reinforced and augmented, or allowed to whither, throughout life? Did the narrowness of erotic thoughts attenuate the channels on which these thoughts travel within the brain, thin the ranks of neurotransmitters that flash along these paths, lead, in turn, to more constriction of thinking? Did the lessons delivered to girls about what is and isn't natural, normal, leave these circuits less sturdy from early on? And broaden opposing tracks, channels of serotonin that rush to quell unacceptable impulse?

"I surreptitiously gaze at him from beneath my lashes as he stands in line waiting to be served. I could watch him all day. He's tall, broad shouldered, and slim, and the way those pants hang from his hips."

Wendy had just read *Fifty Shades of Grey*, the first book in the trilogy of erotica that was approaching, in America, twenty million copies sold, that was breaking records for weekly sales rates, that Wendy and so many others labeled and laughed about as "mommy porn." It wasn't her usual reading. She took in scenes like this as Anastasia, the heroine, recounts the beginnings of her sadomasochistic affair with Christian, his manner reticent and self-possessed, his fingers "graceful," all of him "heart-stoppingly beautiful."

"'Does this mean you're going to make love to me tonight, Christian?'"

"'No, Anastasia, it doesn't. First, I don't make love. I fuck . . . hard.'"

And, soon, like this:

"I come instantly, again and again, falling apart beneath him as he continues to slam deliciously into me."

And, later, like this, with Christian commanding her, "'Hold out your hands in front as if you're praying.' . . . He takes a cable tie and fastens it around my wrists, tightening the plastic. 'Hold onto the post,' he says. . . . He stands behind me and grasps my hips. . . . He smacks me across my behind with his hand. . . . 'Part your legs.' . . . He reaches over me and grabs my braid near the end and winds it around his wrist to my nape, holding my head in place. Very slowly he eases into me, pulling my hair at the same time. . . . His other hand grabs my hip, holding tight, and then he slams into me, jolting me forward. . . . I grip the post harder. . . . He continues his merciless onslaught. . . . My scalp is getting sore from his tugging my hair. . . . I fear my orgasm. . . . If I come I'll collapse. . . . His breathing harsh . . . slamming really deep . . . my name on his lips. . . . I become all body and spiraling sensation and sweet, sweet release, and then completely and utterly mindless."

While Wendy read on her iPad, a storm knocked out power

on her block for a week; this jumbled her family's routines and left them sleeping at a neighbor's house, so there was no telling, she said, whether the book would have accomplished what Flibanserin and her first set of EB pills didn't, whether some of the quickening it caused would have seeped into her feelings for her husband. That week, her life was too much of a mess. She guessed that *Fifty Shades* would have accomplished something, if the circumstances had been different—maybe not all she hoped for from the drugs, but something.

Meana—as Christian's self-possession gave way to "groaning," "slamming," and as Anastasia, bound, bent, became purely object—lay close. But I was thinking of Pfaus's perspective. "Dopamine, dopamine, dopamine," he said about the book's impact. "*Fifty Shades* is activating the whole neurochemical soup of wanting." For Wendy, it was like a series of injections, lasting hours, into a mind that habitually kept fantasy and its neural effects at bay.

And Pfaus added, "Dendrites." These are the gossamer-like tentacles that link neural fibers in our brains. Our experiences can make these tentacles grow more dense, just as plant life thickens in rich soil, and this flourishing, he explained, means "neural networks are enhanced, more sensitized, more capable of being activated." It was possible to imagine that if, for Wendy, devouring the book led to devouring the trilogy, and if this led to more fantasy, if men on the street with fabulous shoulders and hips induced flares of lust, then, over time, "dendritic arborization" might increase and Wendy might find herself at least a bit more eager for her husband, even if his shoulders weren't as broad and his hips weren't as slim and his fucking wasn't as fierce or as new as Christian's, and her name on his lips didn't bring on vertigo.

"Yes," Adriaan Tuiten said, he thought often about reinforcement and neglect, about the bolstering or weakening of the circuits of desire, as he developed Lybrido and Lybridos. He was the founder of Emotional Brain. He was a Dutch researcher in his late fifties with a doctorate in psychopharmacology, whose shirt collars were skewed, whose hair was rumpled, whose dishevelment was half style and half disarray. We met periodically when he was in New York to check on the trials and to sell partnership rights and raise millions for the studies the FDA would still require. He was putting everything into getting past the American agency before its European equivalent; it was too expensive to do both at once. As we walked through Manhattan or leaned over coffee, he railed sometimes that back in the Netherlands people were rifling through his garbage. International companies, vastly larger than his own outfit of forty, were sending spies to get hold of his secrets. They were hacking into EB's computers. Behind his chunky, tinted glasses, his eyes filled with anxiety. He seemed to be, now and then, on the edge of paranoid, crazed. But how crazy were his suspicions? So much money was at stake. And scientists like Pfaus, whose rats hadn't been enlisted by EB, but who knew the field perhaps better than anyone—who had a small advisory role on Lybrido and Lybridos but no monetary interest in EB's success—said that Tuiten could well be the one.

Yet when Tuiten spoke about the conception of his drugs,

the germ of his ideas, a story of scientific ingenuity and monumental potential profit came down to a young man's broken heart. It wasn't something he wished to recollect. "What I'm working on now is functionally independent of the past," he said. Then, slowly: "The starting point is very personal."

Abruptly, when he was in his mid-twenties, his girlfriend, a woman he had been in love with since the age of thirteen and had lived with for years, told him she was leaving. "I was—flabbergasted. You can say that?" he asked me, making sure, in his stiffly accented, halting, but elaborate English that he was using the right word. "I was shocked. I was suffering. And she told me something at that point. She said she was so relieved by her decision that her menstruation came back." She'd stopped using oral contraceptives two years earlier, but her period didn't return, not until the day after she pronounced the relationship over. She believed her body was confirming that she'd made the right choice, no matter how agonizing it had been.

He felt stricken. But it wasn't long before she asked for another chance, and he took her back. "And after a year, the same pattern repeated." She had started taking the pill again, then quit, then went months without ovulating or menstruating; meanwhile, she realized that she really was not meant to be with this man she'd been entangled with for half her life. She let him know it was absolutely finished. And within a day or two: her period.

Battered by this cosmic verdict, he talked with the woman's sister, who was sympathetic but informed him that yes, of course, there could be emotional causes for long stretches without menstruation, that sadly it all made sense. He wasn't a scientist then. He was a belated university student, who'd been determined, until recently, to be a furniture builder. But he'd gravitated back to school because of books friends had given him, books that had begun to captivate him, volumes by the logician and philosopher Bertrand Russell and by Johannes Linschoten, a Dutch experimental psychologist. His mind was shifting, growing more and more avidly analytic. And it dawned on him that something was amiss. If breaking up had set his soul mate's body free to bleed, how had this occurred within twenty-four or forty-eight hours?

How, he thought, had she skipped past ovulation and the two weeks that generally need to go by? Her uterus couldn't have compressed into a day or two what takes half a month. True, it was conceivable that, both times, she'd resolved to end it two weeks before letting him know, but this wasn't the story she told, and, brooding about how he'd been so blindsided, ended up so shattered, about how it could have happened twice—"I stood always under the shower, thinking and thinking"—he started to reverse his girlfriend's logic, the logic her sister affirmed.

The reversal began as an insight, a glimmer, and slowly gained solidity as he scoured obscure journals, studying anything that might be at least tangentially related to his notion. "I'm a little bit—not insane. But. There became a need for me to understand my personal life in this way, to have a theory, an instrument, to have control."

His girlfriend was a runner, a vegetarian, a dieter. This was a recipe for amenorrhea, the ceasing of the menstrual cycle—a recipe that was little researched in those days. Her regimen had inflicted havoc on her hormonal system and delayed the renewal of her period after she gave up the pill, he felt sure as he read what he could find. It had also reconfigured "her affective life," he said while we talked in a café, the remote and professional word "affective" at odds with his expression, his voice. Thirty years afterward, he was wistful, bereft. For a long while, with her amenorrhea ravaging her hormones and her brain's biochemistry, she'd lost desire for him. And with desire, her love, too, had flattened. But eventually, with the pill out of the picture and probably with a slight, unnoted easing of her running, her diet, hormones had revived, ovulation had resumed. In the brain, the molecules of eros had resurged. This didn't restore her emotions for him, though. Instead, her reawakened sex drive seemed to flee straight from him toward the wish for other men. "The biological changed her affective feelings for me," he used the word again, scientific, crushed. She decided to cut the bond, to wrest herself free. The sudden switch in her molecular state had cost him the love of his life.

Both times, coincidentally, it had taken her about two weeks to make the decision. His girlfriend and her sister had got it wrong, he thought. The psychological hadn't dictated the hormonal. Rather, biochemistry had determined the trajectory of lust and love; it had destroyed everything.

Tuiten's reasoning had led to his writing and publishing his first scientific paper while he was still working toward his master's (the article was about discerning causality, he said, but "nothing about my suffering"), to his seeking his PhD, to his researching a biochemical vortex that can suck girls into anorexia, to his studying sex. Throughout it all, there were themes, threads that converged in his current inventing. One was the reign of the chemical within the psychological; another was timing. There was the molecular narrative, the biochemical chronology, behind his own tragedy. There was his resequencing of the forces that pull girls into anorexia. There was, after an experience with another girlfriend, his scrutiny of the molecular relationships that plunge certain women into severe premenstrual syndrome, with deprivations of serotonin and, for some, lowered inhibition and crests of lust. There was, much later, his investigation of the exact delay between doses of testosterone and the spurring of desire in women who respond to infusions of the hormone. There were his ruminations on the timing of serotonin pulses and on the timing of compounds that temporarily suppress this neurotransmitter.

Why were some women more prone than others to have desire plummet for their long-term partners, as habit and entrenched commitment robbed spark from stimuli? Why were some less or better able to feel a moderate flame? Why were a few capable of decades and decades of combustion, thrall? Baseline measures of blood-borne hormone weren't much of a predictor, but Tuiten and his EB experts examined how efficiently a woman's brain cells guided the testosterone molecule through cell interiors, so the hormone could do its transformative work, setting off the chemical changes that prime the erotic. Cells that do this guiding in a begrudging way—with molecular receptors that are resistant—might make a plentiful amount of free-floating testosterone partially irrelevant. Welcoming receptors could help a woman do a lot with a minimal quantity. One strand in the weblike thinking behind Tuiten's drugs was spun from genetic coding that hinted at the character of those receptors. Blood could be read for that coding; the personality of the receptors could be deduced. This was one element in EB's effort to peer into the molecular components of the sexual psyches of individual women. It was one reason why Goldstein saw the company's work as a breakthrough and as a possible answer to his testosterone riddles.

Another angle on the testosterone system, on its capacity to incite dopamine, relied on something more crude than genetic coding. It involved measuring the second and fourth fingers of a woman's right and left hands, and calculating the relationship between the index and ring digits. Wendy and all the other EB subjects, when they'd first been interviewed for the trials, had been asked to put their hands on a computer scanner. These images had been sent off to the company. Tuiten was building on emerging evidence, from humans and from rats, that the difference in length between the two fingers was another reflection of how receptive a person's cells were to testosterone in both brain and bones.

Then there was the serotonin network, headquartered toward the front of the brain, a network that can override dopamine, that filters out stimuli and subdues urges, that is responsible for keeping us calm, rational, organized. For a glimpse of serotonin's wiring, Tuiten gazed at another genetic script, illumined by using a fluorescent dye, an electrified gel.

But that was all to do with the inborn. He incorporated, too, as best he could, the learned. He knew that the social impact on the skeins of serotonin and dopamine, on their relative health, on the way they collaborated or competed, was crucial. Serotonin could either add the right drop of coherence to the sexual brain or it could interfere, inhibit, shut eros down. He knew that what the culture repressed or rewarded molded these networks. To gauge this, he used a series of questions, dealing with arousal and orgasm and frequency of masturbation. In combination, the answers spoke—imperfectly, tellingly—to inhibition's intensity. He placed a woman's replies, as well as her genetic coding and finger ratios, into an equation, an algorithm. The equation used eleven elements in all. In this way, he pieced together a vision of a woman's erotic neurology.

This could sound like lunacy. But it was the most detailed attempt at comprehension by any drug company so far. It fed into the composition of his two medicines and into EB's sorting of which women should take which one. The drugs were to be taken a few hours before a woman wanted to feel overwhelmed by eros. Each drug consisted of two parts: a peppermint-flavored coating of testosterone melted in the mouth; an inner pill was swallowed when the peppermint faded away.

In Lybrido, the pill was a cousin of Viagra. In Lybridos, it was a compound called buspirone. And here Tuiten's long obsession with timing was at work. He'd realized that he could arrange a meeting, so that testosterone's peak hours of sexual priming would coincide with the aid many women would need from the other two chemicals. This help, in the case of the Viagra-like chemical, was a heightening of genital swelling, which ramped up sensation and triggered the brain to produce more dopamine. In the case of buspirone, it was a squelching of serotonin. In their different ways, both Lybrido and Lybridos altered the interplay between serotonin and dopamine.

Lybridos was perhaps the more intriguing invention, the clearer example of Tuiten's fixation on timing. Buspirone is an antidepressant. And like all antidepressants, it elevates serotonin. But there is a distinction. Unlike the most popular compounds for depression, the selective serotonin reuptake inhibitors, the SSRIs, buspirone causes, at first, a brief slow-down in the release of the neurotransmitter. And if buspirone isn't taken every day, the gradual rise in serotonin won't occur. The critical effect is serotonin's very short-term suppression. Put this together with testosterone's key hours of stoking dopamine and, even if that stoking was half-crippled by begrudging receptors, Tuiten might provoke an interval of lust; he might provide a replica of what had been felt long ago, when none of this manipulation had been necessary—when the newness of a lover had sent the biochemicals of desire into a frenzy.

It seemed that Tuiten, disheveled and eternally heartbroken, was about to be astonishingly rich. A minor yet enormous reason was this: over fifteen million American women, and countless more around the world, depend on SSRIs to battle their melancholy. Some would be enrolled in an upcoming phase of the trials. With the boosting of serotonin brought about by SSRIs, there is, for most, an inevitable flight of eros. Desire can become dim, imperceptible. And this can be worsened by the fact that excesses of serotonin disrupt the physical mechanics of orgasm, impeding contractions, so that climax feels farther and farther off, until it is unattainable. For women on SSRIs, each tablet of Lybridos, with its temporary blocking of serotonin, would grant eros a reprieve.

But above all, if Tuiten's thinking was right, if the data he already held, from small sets of women in provisional studies, were borne out in his larger trials, then he had conjured a pair of drugs that were an antidote to monogamy. His medicine promised to rescind years.

Lust, in its most powerful moments, can propel us outside ourselves, outside the world, outside time. We are offered this oblivion. How wonderful the trance is—or was, if such moments have been sacrificed, lost in the quest for another kind of escape: for safety, for constancy, for a fortification against growing old alone, against enduring the terrors of time by ourselves. Could Tuiten's pills perform a type of magic, allowing the trance to co-exist with the comfort? Permitting both kinds of escape? Could Tuiten's chemicals execute that trick?

Wendy prayed that her first round of EB tablets had been placebos. But if, for her, Tuiten's drugs didn't succeed, she said, "There's got to be something else. They've got all these meds for all these other psychiatric issues. Something's got to pop up that can help with this. Right? Right? Right?"

All she was asking was that the round of pills she was now bringing home would stop and reverse what she called "the dwindling." All she was asking was that the medicine would drive her to take her husband's hand at the base of the stairs and draw them together to the top. All she wanted was this: to leave time inconsequential, to turn time into nothing at all.

CHAPTER TEN

A Beginning

Lifting her arm and ringing a little bell, the monitor calls out, "Men rotate! Men rotate!" In the cocktail lounge that's been rented for the evening, each man rises from a small square table, turns from the woman he's been talking with, and steps toward his next assignment. The women wait. They sit along a low, L-shaped banquette. In a pink blouse with a ruffled neckline, in a tight, black cardigan, in a dress with sleeves of gauze, they stay where they are, folded into the maroon upholstery, gazing upward to find out who will appear before them. For a few seconds, the men stride through the soft glow.

This is speed dating. The dates last four minutes, marked by the high-pitched bell. At the session's end, all the women and men submit their decisions privately to the speed-dating company—a yes or no on each of the ten people they met, an expression of interest or an expression of none. Any pair who said yes to each other is put in contact.

The setting isn't always a lounge. The bell is sometimes a playful gong, sometimes just a command. Four minutes is sometimes eight, sometimes three. But one aspect rarely varies: the men move, step near, before taking the seat opposite; the women remain still. The companies explain the convention by observing that women have handbags and that switching spots for them would take more time. Or they note expectations: that men should make the symbolic gesture of chivalry, getting up from their chairs and taking the initiative, while the women need only perch comfortably. This is just the way it is.

And since speed dating caught on in America and Europe after its invention in the late nineties by a Los Angeles rabbi desperate to make Jewish matches, researchers have used the form to examine patterns of desire. They've studied the statistics of a company named Hurrydate, tallying the choices of ten thousand clients. They've created evenings of their own, following all the speed-dating traditions and compiling their own numbers. And again and again, a contrast has emerged: when it comes to wanting a second date, a real date, women are far more selective than men, far less likely to say yes.

For evolutionary psychologists, this has added confirmation to certainties already established. Men are programmed to pursue and inseminate, pursue and inseminate, women to choose the just-right mate. Genetically, men are designed to lust wildly, women to desire in distinct moderation.

But two psychologists, Eli Finkel at Northwestern University and Paul Eastwick at the University of Texas at Austin, noticed what is known to scientists as a confound, a factor that might distort the data, insinuating illusion under the guise of insight. The factor was obvious, yet none of the speed-dating researchers made anything of it. No one discussed it in their academic papers; no one treated it as relevant. What would happen, Finkel and Eastwick wondered, if the instruction was "Women rotate," if the men waited while the women stood and strode forward?

The science and thinking I have brought together in this book are a beginning, only that. None of the researchers I have learned from, not Meredith Chivers or Kim Wallen, not Marta Meana or Jim Pfaus, would claim to have definitive, fully formed answers about female desire. All of them, no matter how evocative their experiments and piercing their ideas, are acutely aware of the layers of unknowns—and of the impediments to getting beneath. The investigation of women's sexual psyches is, with the exception of pharmaceutical quests, dismally funded, supported in strangely inverse proportion to its importance. Eros lies at the heart of who we are as human beings, yet we shun the study of our essential core, shun it perhaps most of all where it is least understood, in women. Where there should be an abundance of exploration, there is, instead, common assumption, unproven theory, political constraint, varieties of blindness.

Once, I asked Chivers why I never found myself phoning the psychology departments of Harvard or Yale or Princeton, why I never spent time with their professors, why so few of America's most elite universities devoted any attention to her field. "Because there is a kind of taboo," she said. "Because we who do this work are second-class citizens." Second-class citizens for digging toward the primary, the primitive, the primal. Unseemly to be down there, metaphorically, literally. And unsettling to have scientists constantly threatening to send back information that might, experiment by experiment, study by study, paper by paper, tear presumption to shreds.

The presumption that while male lust belongs to the animal realm, female sexuality tends naturally toward the civilized; the belief that in women's brains the more advanced regions, the domains of forethought and self-control, are built by heredity to ably quiet the libido; the premise that emotional bonding is, for women, a potent and ancestrally prepared aphrodisiac; the idea that female eros makes women the preordained if imperfect guardians of monogamy—what nascent truths will come into view, floating forward if these faiths continue to be cut apart?

Finkel and Eastwick set up fifteen speed-dating events with a total of three hundred and fifty women and men. At half of the gatherings, the men carried out the approaches. At the rest, when the bell sounded, the women took this part; in just this one momentary way, repeatedly over the course of an hour, traditional romantic roles were upended. A hint of Deidrah, of the sexually stalking rhesus females, was written into the rules.

The researchers asked the participants not only to check yes or no after each four-minute meeting but to rate their sexual feelings for every partner.

The results were straightforward. Social structure—and maybe something imbedded physically in the act of initiating—altered perceptions, decisions, eros. Improbably, yet unmistakably, the shift took hold right away. The numbers were plain. When the women were the ones who moved near, they said yes as often, as indiscriminately, as the men. When the women were the ones who crisscrossed the room and closed in, their ratings of desire became just as lustful. With the rules adjusted, a new reality leapt fleetingly into sight.

READINGS

Behind this book lies a labyrinth of reading. There are the scores of books that line my shelves, from Richard Posner's cost-benefit analysis of erotic motivation, *Sex and Reason*, to Karen Horney's reappraisal of Freud, *Feminine Psychology*, from a collection of sexologists' biographies, *How I Got into Sex*, to Max Wolf Valerio's memoir of metamorphosis from woman to man, *The Testosterone Files*, to a legion of sexual self-help volumes spanning the pragmatic and the spiritual. In the following list of readings, I include a few of the books that my readers might find most directly relevant to the topics I've raised, as well as academic papers that detail much of the research I've written about (though what I've learned from these papers has been dwarfed by what I've taken in through conversations with researchers) and whose footnotes will offer a beginning to anyone who wants to enter the maze of sexual science I've lived in for the past eight years.

I start with Meredith Chivers, whose work is discussed in chapters one, two, and six. (Always scrupulous—at once the bold sexologist and the careful statistician—she asked me to note that the comparison of responses to strangers and close friends in chapter two relies on standard deviations as opposed to absolute values.) Her relevant papers, in order of publication date, are:

Chivers, M. L., & Timmers, A. D. (2012). The effects of gender and relationship context cues in audio narratives on heterosexual women's and men's genital and subjective sexual response. *Archives of Sexual Behavior, 41*, 187–197.

Chivers, M. L., Seto, M. C., Lalumiére, M. L., Laan, E., & Grimbos, T. (2010). Agreement of genital and subjective measures of sexual arousal in men and women: a meta-analysis. *Archives of Sexual Behavior, 39*, 5–56.

Suschinsky, K., Lalumiére, M. L., & Chivers, M. L. (2009). Sex differences in patterns of genital arousal: measurement artifact or true phenomenon? *Archives of Sexual Behavior, 38*, 559–573.

Chivers, M. L., Seto, M. C., & Blanchard, R. (2007). Gender and sexual orientation differences in sexual response to the sexual activities versus the gender of actors in sexual films. *Journal of Personality and Social Psychology, 93*, 1108–1121.

Chivers, M. L., & Bailey, J. M. (2005). A sex difference in features that elicit genital response. *Biological Psychology, 70*, 115–120.

Chivers, M. L., Rieger, G., Latty, E., & Bailey, J. M. (2004). A sex difference in the specificity of sexual arousal. *Psychological Science, 15*, 736–744.

Studies by Terri Fisher and by Terri Conley appear in chapter two; they are:

Alexander, M. G., Fisher, T. D. (2003). Truth and consequences: using the bogus pipeline to examine sex differences in self-reported sexuality. *Journal of Sex Research, 40*, 27–35.

Fisher, T. D. (in press). Gender roles and pressure to be

truthful: the bogus pipeline modifies gender differences in sexual but not non-sexual behavior. *Sex Roles*.

Conley, T. D. (2011). Perceived proposer personality characteristics and gender differences in acceptance of casual sex offers. *Journal of Personality and Social Psychology, 100*, 309–329.

Turning to chapter three, for further study of the history of female sexuality since classical times—or, rather, of the way female sexuality has been perceived—the scholarship of Thomas Laqueur may be the best place to begin:

Laqueur, T. (1990). *Making sex: body and gender from the Greeks to Freud*. Cambridge, MA: Harvard University Press.

In his exploration of sexual and societal transformations of the seventeenth and eighteenth centuries, Faramerz Dabhoiwala delves into a wide range of cultural factors that contributed to women being viewed, in the nineteenth, twentieth, and early twenty-first centuries, as the less libidinous gender:

Dabhoiwala, F. (2012). *The origins of sex: a history of the first sexual revolution*. New York: Oxford University Press.

Nancy Cott provides an analysis of Victorian perspectives:

Cott, N. (1978). Passionlessness: an interpretation of Victorian sexual ideology, 1790–1850. *Signs: Journal of Women in Culture and Society, 4*, 219–236.

The work of David Buss is central to evolutionary psychology's view of human sexuality, and Louann Brizendine offers a popular primer:

Buss, D. M. (1995). *The evolution of desire: strategies of human mating.* New York: Basic Books.

Buss, D. M., & Schmitt, D. P. (1993). Sexual strategies theory: an evolutionary perspective on human mating. *Psychological Review, 100*, 204–232.

Brizendine, L. (2006). *The female brain*. New York: Broadway Books.

The health education programs quoted in chapter three are from curricula produced by Choosing the Best Publishing of Atlanta, Georgia, and by the Center for Relationship Education of Denver, Colorado. Each organization has recently altered some of its language, but the curricula continue to include declamations like "Men respond sexually by what they see and women respond sexually by what they hear and how they feel about it."

Chapter four is devoted primarily to the research of Kim Wallen and Jim Pfaus, and Pfaus in turn emphasizes the importance of experiments conducted by Raul Paredes:

Wallen, K., & Rupp, H. A. (2010). Women's interest in visual sexual stimuli varies with menstrual cycle phase at first exposure and predicts later interest. *Hormones and Behavior, 57*, 263–268.

Rupp, H. A., & Wallen, K. (2007). Sex differences in viewing sexual stimuli: an eye-tracking study in men and women. *Hormones and Behavior, 51*, 524–533.

Wallen, K. (2000). Risky business: social context and hormonal modulation of primate sexual desire. In K. Wallen & J. Schneider (Eds.), *Reproduction in context: social and environmental influences on reproductive physiology and behavior* (pp. 289–323). Cambridge, MA: MIT Press.

Wallen, K. (1990). Desire and ability: hormones and the regulation of female sexual behavior. *Neuroscience and Biobehavioral Reviews, 14*, 233–241.

Wallen, K. (1982). Influence of female hormonal state on rhesus sexual behavior varies with space for social interaction. *Science, 217*, 375–377.

Pfaus, J. G., Kippin, T. E., Coria-Avila, G. A., Gelez, H., Afonso, V. M., Ismail, N., & Parada, M. (2012). Who, what, where, when (and maybe even why): how the experience of sexual reward connects sexual desire, preference, and performance. *Archives of Sexual Behavior, 41*, 31–62.

Georgiadis, J. R., Kringelbach, M. L., & Pfaus, J. G. (2012). Sex for fun: a synthesis of human an animal neurobiology. *Nature Reviews Urology*, *9*, 486–498.

Pfaus, J. G., Wilkins, M. F., DiPietro, N., Benibgui, M., Toledano, R., Rowe, A., & Crouch, M. C. (2010). Inhibitory and disinhibitory effects of psychomotor stimulants and depressants on the sexual behavior of male and female rats. *Hormones and Behavior, 58*, 163–176.

Pfaus, J. G. (2009). Pathways of sexual desire. *Journal of Sexual Medicine, 6*, 1506–1533.

Pfaus, J. G., Giuliano, Francois, & Gelez, H. (2007). Bremelanotide: an overview of preclinical CNS effects on female sexual function. *Journal of Sexual Medicine*, *4*, 269–279.

Martinez, I., & Paredes, R. G. (2001). Only self-paced mating is rewarding in rats of both sexes. *Hormones and Behavior, 40*, 510–517.

Paredes, R. G., & Vasquez, B. (1999). What do female rats like about sex? Paced mating. *Behavioural Brain Research, 105*, 117–127.

The narcissistic element in female desire, the prevalence of rape fantasies, as well as other subjects that come up in chapters five and six are explored in:

Sims, K. E., & Meana, M. (2010). Why did passion wane? A qualitative study of married women's attributions for declines in sexual desire. *Journal of Sex and Marital Therapy, 36*, 360–380.

Lykins, A. D., Meana, M., & Strauss G. P. (2008). Sex differences in visual attention to erotic and non-erotic stimuli. *Archives of Sexual Behavior, 37*, 219–228.

Young-Bruehl, E. (Ed.) (1990). *Freud on women: a reader.* New York: W. W. Norton.

Klein, M. (1975). *Envy and gratitude and other works, 1946–1963.* New York: Delacorte Press/S. Lawrence.

Critelli, J. W., & Bivona, J. M. (2008). Women's erotic rape fantasies: an evaluation of theory and research. *Journal of Sex Research, 1*, 57–70.

Meston, C. M., & Frohlich, P. F. (2003). Love at first fright: partner salience moderates roller-coaster induced excitation transfer. *Archives of Sexual Behavior, 32*, 537–544.

Fedoroff, J. P., Fishell, A., & Fedoroff, B. (1999). A case series of women evaluated for paraphilic disorders. *The Canadian Journal of Human Sexuality, 8*, 127–140.

My discussion of monogamy in chapter seven focuses partly on the *Diagnostic and Statistical Manual of Mental Disorders*, and at the time of my writing the current *DSM* was the Fourth Edition, Text Revision, Arlington, VA: American Psychiatric Association. The Fifth Edition (the *DSM-V*) is due to be published in 2013. To fully understand the magnitude of the changes regarding female desire that are being incorporated in this upcoming volume, it would be necessary to study editions going back at least as far as the *DSM-III* of 1980. But one representative detail is the substitution, in the new version, of the phrase "sexual interest" for the phrase "sexual desire." In this and other ways, Basson's vision of cognitive, unlustful decisions, as opposed to erotic drive, is being codified as the female norm. For a complete discussion of the *DSM-V* language and the rationale behind it, read Brotto, L. A. (2010). The *DSM* diagnostic criteria for hypoactive sexual desire disorder in women. *Archives of Sexual Behavior, 39*, 221–239.

The following work also lends vantage points on the concerns of chapter seven:

Basson, R. (2003). Biopsychosocial models of women's sexual response: applications to management of "desire disorders." *Sexual and Relationship Therapy, 18*, 107–115.

Basson, R. (2000). The female sexual response: a different model. *Journal of Sex and Marital Therapy, 26*, 51–65.

Brotto, L. A., Erskine, Y., Carey, M., Ehlen, T., Finlayson, S., Heywood, M., Kwon, J., McAlpine, J., Stuart, G., Thomson, S., & Miller, D. (2012). A brief mindfulness-based cognitive behavioral intervention improves sexual functioning versus wait-list control in women treated for gynecological cancer. *Gynecological Oncology, 125*, 320–325.

Brotto, L. A., Basson, R., & Luria, M. (2008). A mindfulness-based group psychoeducational intervention targeting sexual arousal disorder in women. *Journal of Sexual Medicine, 5*, 1646–1659.

Brotto, L. A., Heiman, J. R., Goff, B., Greer, B., Lentz, G. M., Swisher, E., Tamimi, H., & Blaricom, A. V. (2008). A psychoeducational intervention for sexual dysfunction in women with gynecological cancer. *Archives of Sexual Behavior, 37*, 317–329.

Hrdy, S. B. (2000). The optimal number of fathers: evolution, demography, and history in the shaping of female mate preferences. *Annals of New York Academy of Sciences*, *907*, 75–96.

Hrdy, S. B. (1997). Raising Darwin's consciousness: female sexuality and the prehominid origins of patriarchy. *Human Nature*, *8*, 1–49.

Hrdy, S. B. (1981). *The Woman That Never Evolved*. Cambridge, MA: Harvard University Press.

Hrdy, S. B. (1979). Infanticide among animals: a review, classification, and examination of the implications for the reproductive strategies of females. *Ethology and Sociobiology*, *1*, 13–40.

Zeh, J. A., Newcomer, S. D., & Zeh, D. W. (1998). Polyandrous females discriminate against previous mates. *Proceedings of the National Academy of Sciences*, *95*, 13732–13736.

Diamond, L. M. (2008). *Sexual Fluidity: understanding women's love and desire*. Cambridge, MA: Harvard University Press.

On the disputes over varieties of female orgasm—covered in chapter eight—discussions held in the pages of the *Journal of Sexual Medicine* are a useful way to begin:

Jannini, E. A., Rubio-Casillas, A., Whipple, B., Buisson, O., Komisaruk, B. R., & Brody, S. (2012). Female orgasm(s): one, two, several. *Journal of Sexual Medicine*, *9*, 956–965.

And then there's Barry Komisaruk's and Beverly Whipple's more pragmatic guide to the science of climaxing:

Komisaruk, B. R., Beyer-Flores, C., & Whipple, B. (2006). *The Science of Orgasm*. Baltimore: Johns Hopkins University Press.

For the recent history of female desire drugs chronicled in chapter nine, I've drawn mostly from innumerable conversations with experts in the field, but the mainstream press has written extensively about these failures, and a Web search of the drugs' names, from Intrinsa to Bremelanotide, from Flibanserin to Libigel, will turn up a wealth of further reading.

Finally, about speed dating:

Finkel, E. J., & Eastwick, P. W. (2009). Arbitrary social norms influence sex differences in romantic selectivity. *Psychological Science, 20*, 1290–1295.

ACKNOWLEDGMENTS

Without the voices of the many women who spoke with me, in confidence, about their erotic lives, I could not have written this book. My gratitude goes to those whose stories I've told and to all the others whose thoughts have informed my thinking. Equally, I am indebted to the patient teaching of an array of scientists and clinicians. Beyond those noted within the book, Kelly Allers, Monica Day, Ann D'Ercole, Leonard DeRogatis, Muriel Dimen, Katherine Frank, Irwin Goldstein, Bat Sheva Marcus, Margaret Nichols, Adam Safron, Michael Sand, and Claire Yang were especially generous with their time and perspective.

I have been immensely fortunate to have Suzanne Gluck as my agent throughout my writing life—my thanks go to her and to William Morris Endeavor's Eve Attermann, Raffaella De Angelis, Tracy Fisher, and Alicia Gordon.

Lee Boudreaux, my nimble and inexhaustible editor, has been a wonderful guide. And Dan Halpern, along with Tina Andreadis, Tamara Arellano, Rachel Elinsky, Mark Ferguson, Erin Gorham, Georgia Maas, Karen Maine, Michael McKenzie, Allison Saltzman, Benjamin Tomek, and Craig Young have made me tremendously thankful to have Ecco/HarperCollins as my publisher.

Ilena Silverman, my thoughtful editor at the *New York Times Magazine*, gave this book its start, and Hanna Rosin and Slate's Double X assisted with a blog early on.

My friends Samantha Gillison, John Gulla, William Hogeland, George Packer, Ayesha Pande, Roland Kelts, Elizabeth Rubin, Laura Secor, Anne Sikora, and Tom Watson were always there to contribute counsel, lend humor, and infuse endurance.

My father continues to be true north for me—no words of thanks can capture my feelings.

Nancy Northup, the mother of my children and my former wife, offered faith for many years and, furthermore, does the kind of political and legal advocacy that makes this book's psychological and intellectual explorations possible.

And my children, Natalie and Miles, who are no longer children, are the minds to which I must measure up and the heartbeat that sustains me.